3-D Geometry: Volume

Containers and Cubes

Grade 5

Also appropriate for Grade 6

Michael T. Battista
Mary Berle-Carman

Developed at TERC, Cambridge, Massachusetts

Dale Seymour Publications®

The *Investigations* curriculum was developed at TERC (formerly
Technical Education Research Centers) in collaboration with Kent State
University and the State University of New York at Buffalo. The work
was supported in part by National Science Foundation Grant No. MDR-
9050210. TERC is a nonprofit company working to improve mathematics
and science education. TERC is located at 2067 Massachusetts Avenue,
Cambridge, MA 02140.

This project was supported, in part,
by the
National Science Foundation
Opinions expressed are those of the authors
and not necessarily those of the Foundation

This book is published by Dale Seymour Publications®, an imprint of the
Alternative Publishing Group of Addison-Wesley Publishing Company.

Managing Editor: Catherine Anderson
Series Editor: Beverly Cory
Consulting Editor: Priscilla Cox Samii
ESL Consultant: Nancy Sokol Green
Production/Manufacturing Director: Janet Yearian
Production/Manufacturing Coordinator: Barbara Atmore
Design Manager: Jeff Kelly
Design: Don Taka
Illustrations: Susan Jaekel, Carl Yoshihara
Cover: Bay Graphics
Composition: Publishing Support Services

Printed on Recycled Paper

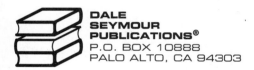

**DALE
SEYMOUR
PUBLICATIONS®**
P.O. BOX 10888
PALO ALTO, CA 94303

Order number DS21429
ISBN 0-86651-993-9
1 2 3 4 5 6 7 8 9 10-ML-99 98 97 96 95

T E R C

INVESTIGATIONS IN NUMBER, DATA, AND SPACE

Principal Investigator   Susan Jo Russell

Co-Principal Investigator   Cornelia C. Tierney

Director of Research and Evaluation   Jan Mokros

Curriculum Development

Joan Akers
Michael T. Battista
Mary Berle-Carman
Douglas H. Clements
Karen Economopoulos
Claryce Evans
Marlene Kliman
Cliff Konold
Jan Mokros
Megan Murray
Ricardo Nemirovsky
Tracy Noble
Andee Rubin
Susan Jo Russell
Margie Singer
Cornelia C. Tierney

Evaluation and Assessment

Mary Berle-Carman
Jan Mokros
Andee Rubin
Tracey Wright

Teacher Support

Kabba Colley
Karen Economopoulos
Anne Goodrow
Nancy Ishihara
Liana Laughlin
Jerrie Moffett
Megan Murray
Margie Singer
Dewi Win
Virginia Woolley
Tracey Wright
Lisa Yaffee

Administration and Production

Irene Baker
Amy Catlin
Amy Taber

Cooperating Classrooms for This Unit

Linda Hallenbeck
Hudson Local School District, Hudson, OH
Alice Madio
Winchester Public Schools, Winchester, MA
Mary Moessner
Bedford Public Schools, Bedford, OH
Sarah Novogrodsky
Cambridge Public Schools, Cambridge, MA

Technology Development

Douglas H. Clements
Julie Sarama

Video Production

David A. Smith
Judy Storeygard

Consultants and Advisors

Deborah Lowenberg Ball
Marilyn Burns
Mary Johnson
James J. Kaput
Mary M. Lindquist
Leslie P. Steffe
Grayson Wheatley

Graduate Assistants

Kent State University:
Richard Aistrope, Kathryn Battista,
Caroline Borrow, William Hunt

State University of New York at Buffalo:
Jeffery Barrett, Julie Sarama,
Sudha Swaminathan, Elaine Vukelic

Harvard Graduate School of Education:
Dan Gillette, Irene Hall

CONTENTS

Teacher Notes

Investigations in Number, Data, and Space is a K–5 mathematics curriculum with four major goals:

- to offer students meaningful mathematical problems
- to emphasize depth in mathematical thinking rather than superficial exposure to a series of fragmented topics
- to communicate mathematics content and pedagogy to teachers
- to substantially expand the pool of mathematically literate students

The *Investigations* curriculum embodies an approach radically different from the traditional textbook-based curriculum. At each grade level, it consists of a set of separate units, each offering 2–6 weeks of work. These units are presented through investigations that involve students in the exploration of major mathematical ideas.

Approaching the mathematics content through investigations helps students develop flexibility and confidence in approaching problems, fluency in using mathematical skills and tools to solve problems, and proficiency in evaluating their solutions. Students also build a repertoire of ways to communicate about their mathematical thinking, while their enjoyment and appreciation of mathematics grows.

The investigations are carefully designed to invite all students into mathematics—girls and boys, diverse cultural, ethnic, and language groups, and students with different strengths and interests. Problem contexts often call on students to share experiences from their family, culture, or community. The curriculum eliminates barriers—such as work in isolation from peers, or emphasis on speed and memorization—that exclude some students from participating successfully in mathematics. The following aspects of the curriculum ensure that all students are included in significant mathematics learning:

- Students spend time exploring problems in depth.
- They find more than one solution to many of the problems they work on.

- They invent their own strategies and approaches, rather than relying on memorized procedures.
- They choose from a variety of concrete materials and appropriate technology, including calculators, as a natural part of their everyday mathematical work.
- They express their mathematical thinking through drawing, writing, and talking.
- They work in a variety of groupings—as a whole class, individually, in pairs, and in small groups.
- They move around the classroom as they explore the mathematics in their environment and talk with their peers.

While reading and other language activities are typically given a great deal of time and emphasis in elementary classrooms, mathematics often does not get the time it needs. If students are to experience mathematics in depth, they must have enough time to become engaged in real mathematical problems. We believe that a minimum of five hours of mathematics classroom time a week—about an hour a day—is critical at the elementary level. The plan and pacing of the *Investigations* curriculum is based on that belief.

For further information about the pedagogy and principles that underlie these investigations, see the Teacher Notes throughout the units and the following books:

- *Implementing the* Investigations in Number, Data, and Space™ *Curriculum*
- *Beyond Arithmetic: Changing Mathematics in the Elementary Classroom*

The *Investigations* curriculum is presented through a series of teacher books, one for each unit of study. These books not only provide a complete mathematics curriculum for your students, they offer materials to support your own professional development. You, the teacher, are the person who will make this curriculum come alive in the classroom; the book for each unit is your main support system.

While reproducible resources for students are provided, the curriculum does not include student books. Students work actively with objects and experiences in their own environment and with a variety of manipulative materials and technology, rather than with workbooks and worksheets filled with problems. We also make extensive use of the overhead projector as a way to present problems, to focus group discussion, and to help students share ideas and strategies. If an overhead projector is available, we urge you to try it as suggested in the investigations.

Ultimately, every teacher will use these investigations in ways that make sense for his or her partic-

ular style, the particular group of students, and the constraints and supports of a particular school environment. We have tried to provide with each unit the best information and guidance for a wide variety of situations, drawn from our collaborations with many teachers and students over many years. Our goal in this book is to help you, as a professional educator, implement this mathematics curriculum in a way that will give all your students access to mathematical power.

Investigation Format

The opening two pages of each investigation help you get ready for the student work that follows. Here you will read:

What Happens—a synopsis of each session or block of sessions.

Mathematical Emphasis—the most important ideas and processes students will encounter in this investigation.

What to Plan Ahead of Time—materials to gather, student sheets to duplicate, transparencies to make, and anything else you need to do before starting.

Packing Problems

What Happens

Sessions 1 and 2: Packing Packages of Different Sizes Student pairs find ways to accurately predict, then determine how many packages of different sizes will fit in a given box. They check their work by making the box and filling it with packages of cubes.

Sessions 3 and 4: Designing Boxes Students design a single box that can be completely filled with each of four or five different-shaped rectangular packages. They pack the box with only one type of package at a time, and it must fill the box to the top with no gaps.

Session 5: More Packing Problems Students predict how many two-cube packages fit into boxes that are marked off by rectangles, not squares. This activity uncovers a misconception many students have about multiplication and arrays.

Mathematical Emphasis

■ Organizing rectangular packages so that they fill rectangular boxes

■ Developing strategies for determining how many rectangular packages fill a box

■ Designing boxes to hold packages of different sizes

■ Understanding the relationship between the dimensions of a box and how many rectangular packages fill the box

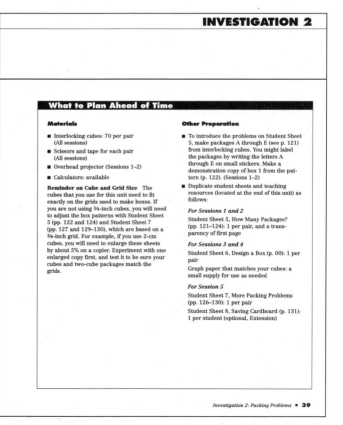

What to Plan Ahead of Time

Materials

■ Interlocking cubes: 70 per pair (All sessions)

■ Scissors and tape for each pair (All sessions)

■ Overhead projector (Sessions 1–2)

■ Calculators: available

Reminder on Cube and Grid Size The cubes that you use for this unit need to fit exactly on the grids used to make boxes. If you are not using ¾-inch cubes, you will need to adjust the box patterns with Student Sheet 5 (pp. 122 and 124) and Student Sheet 7 (pp. 127 and 129–130), which are based on a ¾-inch grid. For example, if you use 2-cm cubes, you will need to enlarge these sheets by about 5% on a copier. Experiment with one enlarged copy first, and test it to be sure your cubes and two-cube packages match the grids.

Other Preparation

■ To introduce the problems on Student Sheet 5, make packages A through E (see p. 121) from interlocking cubes. You might label the packages by writing the letters A through E on small stickers. Make a demonstration copy of box 1 from the pattern (p. 122). (Sessions 1–2)

■ Duplicate student sheets and teaching resources (located at the end of this unit) as follows:

For Sessions 1 and 2
Student Sheet 5, How Many Packages? (pp. 121–124): 1 per pair, and a transparency of first page

For Sessions 3 and 4
Student Sheet 6, Design a Box (p. 00): 1 per pair
Graph paper that matches your cubes: a small supply for use as needed

For Session 5
Student Sheet 7, More Packing Problems (pp. 126–130): 1 per pair
Student Sheet 8, Saving Cardboard (p. 131): 1 per student (optional, Extension)

Sessions Within an investigation, the activities are organized by class session, a session being a one-hour math class. Sessions are numbered consecutively through an investigation. Often several sessions are grouped together, presenting a block of activities with a single major focus.

When you find a block of sessions presented together—for example, Sessions 1, 2, and 3—read through the entire block first to understand the overall flow and sequence of the activities. Make some preliminary decisions about how you will divide the activities into three sessions for your class, based on what you know about your students. You may need to modify your initial plans as you progress through the activities, and you may want to make notes in the margins of the pages as reminders for the next time you use the unit.

Be sure to read the Session Follow-Up section at the end of the session block to see what homework assignments and extensions are suggested as you make your initial plans.

While you may be used to a curriculum that tells you exactly what each class session should cover, we have found that the teacher is in a better position to make these decisions. Each unit is flexible and may be handled somewhat differently by every teacher. While we provide guidance for how many sessions a particular group of activities is likely to need, we want you to be active in determining an appropriate pace and the best transition points for your class.

Ten-Minute Math At the beginning of some sessions, you will find Ten-Minute Math activities. These are designed to be used in tandem with the investigations, but not during the math hour. Rather, we hope you will do them whenever you have a spare 10 minutes—maybe before lunch or recess, or at the end of the day.

Ten-Minute Math offers practice in key concepts, but not always those being covered in the unit. For example, in a unit on using data, Ten-Minute Math might revisit geometric activities done earlier in the year. Complete directions for the suggested activities are included at the end of each unit. A compilation of Ten-Minute Math activities is also available as a separate book.

Sessions 1 and 2

Packing Packages of Different Sizes

Materials

- Prepared packages A through E and box 1, for demonstration
- Student Sheet 5 (1 per pair, plus transparency of first page)
- Interlocking cubes (70 per pair)
- Scissors and tape for each pair
- Overhead projector

What Happens

Student pairs find ways to accurately predict, then determine how many packages of different sizes will fit in a given box. They check their work by making the box and filling it with packages of cubes. Student work focuses on:

- organizing rectangular packages to fit in rectangular boxes
- developing strategies for determining how many rectangular packages will fit in a box

Activity

How Many Packages?

In this second investigation, students consider situations in which paper boxes are filled with rectangular packages that are not cubic in shape. Each package is made with several cubes. To avoid confusion, it is important to maintain the distinction between *cubes, packages,* and *boxes,* as explained in Investigation 1 (p. 20). Remind students that as they talk about their work in this unit, they always need to explain their thinking with the boxes, packages, and cubes in hand. Help students do this by setting a clear example.

Distribute to each pair the four-page Student Sheet 5, How Many Packages? If possible, also use an overhead to project a transparency of the first page of the student sheet for everyone to see.

In order to show the bottom of the boxes, the diagrams of boxes 1 and 2 on the student sheet are more complex than earlier diagrams. Having a model of box 1 can clarify the diagram for students. As you introduce the problems on Student Sheet 5, show students your actual cube packages A through E and the demonstration box 1.

Here we are back on the job at the packaging factory. We need to make boxes to ship different quantities of these five packages. *[Display the packages you have made from cubes and show how they correspond to the pictured packages A, B, C, D, and E.]* Your job is to find a way to accurately predict how many of each package will fit in a box before it is made. You will be working with a partner on this job.

Activities The activities include pair and small-group work, individual tasks, and whole-class discussions. In any case, students are seated together, talking and sharing ideas during all work times. Students most often work cooperatively, although each student may record work individually.

Choice Time In some units, some sessions are structured with activity choices. In these cases, students may work simultaneously on different activities focused on the same mathematical ideas. Students choose which activities they want to do, and they cycle through them.

You will need to decide how to set up and introduce these activities and how to let students make their choices. Some teachers present them as station activities, in different parts of the room. Some list the choices on the board as reminders or have students keep their own lists.

Excursions Some of the investigations in this unit include *excursions*—activities that could be omitted without harming the integrity of the unit. This is one way of dealing with the overabundance of fascinating mathematics to be studied—much more than a class has time to explore in any one year. Excursions give you the flexibility to make different choices from year to year. For example, you might do the excursions in this 3-D Geometry unit this year, but another year, try the excursions in another unit.

Tips for the Linguistically Diverse Classroom
At strategic points in each unit, you will find concrete suggestions for simple modifications of the teaching strategies to encourage the participation of all students. Many of these tips offer alternative ways to elicit critical thinking from students at varying levels of English proficiency, as well as from other students who find it difficult to verbalize their thinking.

The tips are supported by suggestions for specific vocabulary work to help ensure that all students can participate fully in the investigations. The Preview for the Linguistically Diverse Classroom (p. 13) lists important words that are assumed as part of the working vocabulary of the unit. Second-language learners will need to become familiar with these words in order to understand the problems and activities they will be doing. These terms can be incorporated into students' second-language work before or during the unit. Activities that can be used to present the words are found in the appendix, Vocabulary Support for Second-Language Learners (p. 109).

In addition, ideas for making connections to students' language and cultures, included on the Preview page, help the class explore the unit's concepts from a multicultural perspective.

Session Follow-Up

Homework Homework is suggested on a regular basis in the grade 5 units. The homework may be used for (1) review and practice of work done in class; (2) preparation for activities coming up in class—for example, collecting data for a class project; and (3) involving and informing family members.

Some units in the *Investigations* curriculum have more homework than others, simply because it makes sense for the mathematics that's going on. Other units rely on manipulatives that most stu-

dents won't have at home, making homework difficult. In any case, homework should always be directly connected to the investigations in the unit, or to work in previous units—never sheets of problems just to keep students busy.

Extensions These follow-up activities are opportunities for some or all students to explore a topic in greater depth or in a different context. They are not designed for "fast" students; mathematics is a multifaceted discipline, and different students will want to go further in different investigations. Look for and encourage the sparks of interest and enthusiasm you see in your students, and use the extensions to help them pursue these interests.

Family Letter A letter that you can send home to students' families is included with the blackline masters for each unit. We want families to be informed about the mathematics work in your classroom; they should be encouraged to participate in and support their child's work. A reminder to send home the letter appears in one of the early investigations. These letters are also available separately in Spanish, Vietnamese, Cantonese, Hmong, and Cambodian.

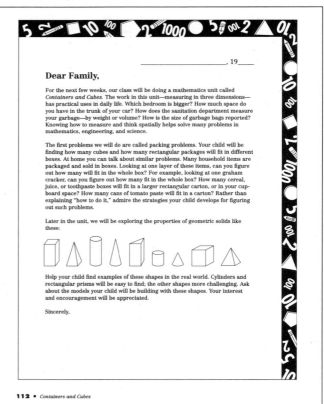

Materials

A complete list of the materials needed for the unit is found on p. 11. Most of these materials are available in a kit for the *Investigations* grade 5 curriculum. Individual items can also be purchased as needed from school supply stores and dealers.

In an active mathematics classroom, certain basic materials should be available at all times: interlocking cubes, pencils, unlined paper, graph paper, calculators, things to count with, and measuring tools. Some activities in this curriculum require scissors and glue sticks or tape. Stick-on notes and large paper are also useful materials throughout.

So that students can independently get what they need at any time, they should know where these materials are kept, how they are stored, and how they are to be returned to the storage area. For example, interlocking cubes are best stored in towers of ten; then, whatever the activity, they should be returned to storage in groups of ten at the end of the hour. You'll find that establishing such routines at the beginning of the year is well worth the time and effort.

Student Sheets and Teaching Resources

Reproducible pages to help you teach the unit are found at the end of this book. These include masters for making overhead transparencies and other teaching tools, as well as student recording sheets.

Many of the field-test teachers requested more sheets to help students record their work, and we have tried to be responsive to this need. At the same time, we think it's important that students find their own ways of organizing and recording their work. They need to learn how to explain their thinking with both drawings and written words, and how to organize their results so someone else can understand them.

To ensure that students get a chance to learn how to represent and organize their own work, we deliberately do not provide student sheets for every activity. We recommend that your students keep a mathematics notebook or folder so that their work, whether on reproducible sheets or their own paper, is always available to them for reference.

Help for You, the Teacher

Because we believe strongly that a new curriculum must help teachers think in new ways about mathematics and about their students' mathematical thinking processes, we have included a great deal of material to help you learn more about both.

About the Mathematics in This Unit This introductory section (page 12) summarizes for you the critical information about the mathematics you will be teaching. This will be particularly valuable to teachers who are accustomed to a traditional textbook-based curriculum.

Teacher Notes These reference notes provide practical information about the mathematics you are teaching and about our experience with how students learn. Many of the notes were written in response to actual questions from teachers, or to discuss important things we saw happening in the field-test classrooms. Some teachers like to read them all before starting the unit, then review them as they come up in particular investigations.

Dialogue Boxes Sample dialogues demonstrate how students typically express their mathematical ideas, what issues and confusions arise in their thinking, and how some teachers have guided class discussions. These dialogues are based on the extensive classroom testing of this curriculum; many are word-for-word transcriptions of recorded class discussions. They are not always easy reading; sometimes it may take some effort to unravel what the students are trying to say. But this is the value of these dialogues; they offer good clues to how your students may develop and express their approaches and strategies, helping you prepare for your own class discussions.

Where to Start You may not have time to read everything the first time you use this unit. As a first-time user, you will likely focus on understanding the activities and working them out with your students. Read completely through each investigation before starting to present it.

When you next teach this same unit, you can begin to read more of the background. Each time you present this unit, you will learn more about how your students understand the mathematical ideas. The first-time user of *Containers and Cubes* should read the following:

- About the Mathematics in This Unit (p. 12)
- Teacher Note: Strategies for Finding the Number of Cubes in 3-D Arrays (p. 26)
- Teacher Note: Strategies for Finding How Many Packages (p. 43)

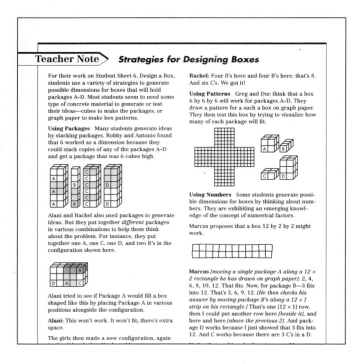

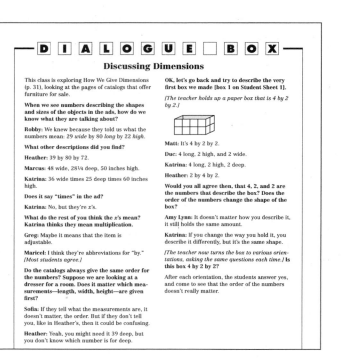

Teacher Checkpoints As a teacher of the *Investigations* curriculum, you observe students daily, listen to their discussions, look carefully at their work, and use this information to guide your teaching. We have designated Teacher Checkpoints as natural times to get an overall sense of how your class is doing in the unit.

The Teacher Checkpoints provide a time for you to pause and reflect on your teaching plan while observing students at work in an activity. These sections offer tips on what you should be looking for and how you might adjust your pacing. Are most students fluent with strategies for solving a particular kind of problem? Are they just starting to formulate good strategies? Or are they still struggling with how to start?

Depending on what you see as the students work, you may want to spend more time on similar problems, change some of the problems to use smaller numbers, move quickly to more challenging material, modify subsequent activities for some students, work on particular ideas with a small group, or pair students who have good strategies with those who are having more difficulty.

In *Containers and Cubes* you will find four Teacher Checkpoints:

Investigation 1, Sessions 1–2:
Writing Up a General Method (p. 23)

Investigation 1, Sessions 3–4:
Halving the Number of Cubes (p. 34)

Investigation 2, Sessions 1–2:
How We Counted Packages (p. 42)

Investigation 4, Sessions 2–3:
Pyramid and Prism Partners (p. 82)

Embedded Assessment Activities Use the built-in assessments included in this unit to help you examine the work of individual students, figure out what it means, and provide feedback. From the students' point of view, the activities you will be using for assessment are no different from any others; they don't look or feel like traditional tests.

These activities sometimes involve writing and reflecting, at other times a brief interaction between student and teacher, and in still other instances the creation and explanation of a product.

In *Containers and Cubes,* you will find two assessment activities:

Investigation 2, Session 5:
More Packing Problems (p. 49)

Investigation 4, Sessions 7–9:
Reviewing Final Reports (p. 101)

Teachers find the hardest part of the assessment to be interpreting their students' work. If you have used a process approach to teaching writing, you will find our mathematics approach familiar. To help with interpretation, we provide guidelines and questions to ask about the student's work. In some cases we include a Teacher Note with specific examples of student work and a commentary on what it indicates. This framework can help you determine how your students are progressing.

As you evaluate students' work, it's important to remember that you're looking for much more than the "right answer." You'll want to know what their strategies are for solving the problem, how well these strategies work, whether they can keep track of and logically organize an approach to the problem, and how they make use of representations and tools to solve the problem.

Ongoing Assessment Good assessment of student work involves a combination of approaches. Some of the things you might do on an ongoing basis include the following:

- **Observation** Circulate around the room to observe students as they work. Ask students to explain their strategies, and listen to their discussions of mathematical ideas.

- **Portfolios** Ask students to document their work, in journals, notebooks, or portfolios. Periodically review this work to see how their mathematical thinking and writing are changing. Some teachers have students keep a notebook or folder for each unit, while others prefer one mathematics notebook or a portfolio of selected work for the entire year. Take time at the end of each unit for students to choose work for their portfolios. You might also have them write about what they've learned in the unit.

Containers and Cubes

Content of This Unit By packing rectangular boxes with cubes, students develop strategies to determine how many cubes or packages fit inside. They explore the concept of volume, inventing strategies for finding the volume of small paper boxes and larger spaces such as their classroom. They investigate volume relationships between cylinders and cones and between pyramids and prisms with the same base and height. They also learn about the structure of geometric solids and improve their visualization skills.

Connections with Other Units If you are doing the full-year *Investigations* curriculum in the suggested sequence for grade 5, this is the eighth of nine units. Through their work in the grade 5 unit *Measurement Benchmarks*, your class will have some experience with measuring the liquid volume of containers in milliliters; here they measure the volume of solids in cubic centimeters. This unit also extends the concepts students encountered in the 3-D Geometry units *Exploring Solids and Boxes* (grade 3) and *Seeing Solids and Silhouettes* (grade 4).

This unit can be used successfully at either grade 5 or grade 6, depending on the previous experience and needs of your students.

Investigations Curriculum ■ Suggested Grade 5 Sequence

Mathematical Thinking at Grade 5 (Introduction and Landmarks in the Number System)

Picturing Polygons (2-D Geometry)

Name That Portion (Fractions, Percents, and Decimals)

Between Never and Always (Probability)

Building on Numbers You Know (Computation and Estimation Strategies)

Measurement Benchmarks (Estimating and Measuring)

Patterns of Change (Tables and Graphs)

▶ *Containers and Cubes* (3-D Geometry: Volume)

Data: Kids, Cats, and Ads (Statistics)

Investigation 1 • The Packaging Factory

Class Sessions	Activities	Pacing	Ten-Minute Math
Sessions 1 and 2 (page 18) HOW MANY CUBES?	How Many Cubes? Discussion: Sharing Ways of Predicting ■ Teacher Checkpoint: Writing Up a General Method ■ Homework ■ Extension	2 hrs	 Counting Around the Class (Decimals)
Sessions 3 and 4 (page 30) DOUBLING AND HALVING BOXES OF CUBES	Checking Homework Predictions How We Give Dimensions Doubling the Number of Cubes ■ Teacher Checkpoint: Halving the Number of Cubes ■ Extensions	2 hrs	

Investigation 2 • Packing Problems

Class Sessions	Activities	Pacing	Ten-Minute Math
Sessions 1and 2 (page 40) PACKING PACKAGES OF DIFFERENT SIZES	How Many Packages? ■ Teacher Checkpoint: How We Counted Packages ■ Extension	2 hrs	 Counting Around the Class (Decimals)
Sessions 3 and 4 (page 45) DESIGNING BOXES	Design a Box ■ Homework	2 hrs	
Session 5 (page 49) MORE PACKING PROBLEMS	Assessment: More Packing Problems ■ Extensions	1 hr	

Investigation 3 • Measuring the Space in Our Classroom

Class Sessions	Activities	Pacing	Ten-Minute Math
Session 1 (page 54) MEASURING THE SPACE IN A BOX	Finding Cubic Centimeters Without Cubes Units of Volume Building Models of Volume Units A Plan to Measure Classroom Space ■ Homework	1 hr	 Counting Around the Class (Decimals)
Session 2 (page 64) THE SPACE INSIDE OUR CLASSROOM	How Many Cubic Meters in Our Classroom? Discussion: How We Measured ■ Homework ■ Extensions	1 hr	
Session 3 (Excursion)* (page 72) MEASURING THE SPACE IN OTHER ROOMS	Measuring Another Room Writing a Report on the Volume Sharing Our Findings	1 hr	

* Excursions can be omitted without harming the integrity or continuity of the unit, but offer good mathematical work if you have time to include them.

Investigation 4 • Prisms and Pyramids, Cylinders and Cones			
Class Sessions	**Activities**	**Pacing**	**Ten-Minute Math**
Session 1 (page 76) COMPARING VOLUMES	Comparing Volumes of Containers Making Solids from Patterns	1 hr	 Guess My Number
Sessions 2 and 3 (page 80) COMPARING VOLUMES OF RELATED SHAPES	Identifying the Different Solids Comparing Solids and Their Volumes Discussion: What We Discovered ■ Teacher Checkpoint: Pyramid and Prism Partners ■ Homework	2 hrs	
Sessions 4 and 5 (page 88) USING STANDARD VOLUME UNITS	Measuring with Cubic Centimeters Discussion: How We Found the Volume Cubic Centimeters in a Liter	2 hrs	
Session 6 (Excursion) (page 93) HOW DO THE HEIGHTS COMPARE?	Comparing Cylinders and Cones Comparing Prisms and Pyramids	1 hr	
Sessions 7, 8, and 9 (page 95) BUILDING MODELS	A Final Project with Geometric Solids Presenting Models to the Class Assessment: Reviewing Final Reports Choosing Student Work to Save ■ Extension	3 hrs	

Following are the basic materials needed for the activities in this unit. Items marked with an asterisk are provided with the *Investigations* Materials Kit for grade 5.

* Interlocking cubes: 70 per pair

* Metersticks: 1 per pair. If metersticks are in short supply, you will need from 3 to 12 sticks in one-meter lengths for building a cubic meter model. Students could use more such sticks or string cut into one-meter lengths for measuring tools.

  Yardsticks: 12 to make a cubic yard model, plus extras if available

  Centimeter rulers: 1 per pair

  Foot rulers: 12 to make a cubic foot model, plus extras if available, or sheets of cardboard larger than 12×12

* Centimeter cubes: 100–150

* See-through graduated prism: 8 per class

* Liter measuring pitchers: 4 per class

  Rice or sand for measuring volume: about 1/2 pound per pair. Rice generally works better, but the sand available from toy stores is a suitable alternative.

  Trays for carrying and containing the rice or sand

  Paper bags: 20–25 paper bags, large enough to hold three small containers

  Furniture catalogs or office or school supply catalogs

  Calculators

  Overhead projector

  Scissors: 1 per pair

  Tape: 1 roll per pair

  Masking tape: 1 roll for each group of 4–6 students

The following materials are provided at the end of this unit as blackline masters. They are also available in classroom sets.

■ Family Letter (p. 112)

■ Student Sheets 1–15 (starting on p. 113)

■ Teaching Resources:

Packaging Factory (p. 118)
More Boxes for Predicting (p. 119)
Large Box (p. 120)
Solid Patterns A through K (pp. 139–143)
Cone, Pyramid, Cylinder Patterns (pp. 144–145)
See-through Graduated Prism Pattern (p. 146)
Dimensions of Solids (p. 147)
Pattern Maker Templates (pp. 148–153)
Three-Quarter-Inch Graph Paper (p. 154)
Two-Centimeter Graph Paper (p. 155)
One-Centimeter Graph Paper (p. 156)
Guess My Unit Cards (pp. 157–158)

Volume is an essential concept in students' learning of three-dimensional or solid geometry. The volume of a solid is the amount of space that an object occupies; it is generally measured with unit cubes. To understand the measurement of volume, students must develop strategies for determining the number of cubes in 3-D arrays by mentally organizing the cubes—for example, as a stack of three rectangular layers, each 3 by 4 cubes.

Students who do not have a proper mental organization for the cubes in a 3-D array often think of such arrays as uncoordinated sets of outside faces. They believe that by counting all or some of the square faces on the outside of an array, they are counting all of its cubes.

In this unit, students make open rectangular boxes from graph paper and find ways to determine how many cubes or "packages" of cubes fit inside. This hands-on activity helps students learn the structure of the rectangular boxes they are trying to measure, as well as the structure of the cube arrays that fit inside. With repeated experiences, students become able to visualize the organization of these cube arrays in the boxes without looking at concrete models. Students use numerical ideas such as skip counting, repeated addition, multiplication, factors, and multiples to solve the box problems.

As they work through the unit, most students will come to determine the number of cubes in rectangular boxes by thinking in terms of layers: "A layer contains 3×4 or 12 cubes, and there are 3 layers, so there are 36 cubes altogether." Traditionally, students have been taught to solve such problems with a formula learned by rote:

Volume = length × width × height

They plug in the numbers and perform the calculation without thinking about why or how the formula works. For meaningful use of the formula, students need to first understand the structure of

3-D arrays of cubes. We strongly discourage teaching this formula to students; the layering strategies that they invent will be more powerful.

Once students develop viable strategies for visualizing and enumerating cubes in 3-D arrays, they apply these strategies to determining the volume of various rectangular containers—from small boxes to classrooms. They then extend their thinking of volume to nonrectangular containers such as pyramids, cylinders, and cones. They find relationships between the volumes of related pyramids and prisms and between the volumes of related cones and cylinders. Furthermore, throughout the unit, students further develop their visualization skills, their knowledge of 3-D figures, and their understanding of relationships between two-dimensional pictures and three-dimensional objects.

Mathematical Emphasis At the beginning of each investigation, the Mathematical Emphasis section tells you what is most important for students to learn about during that investigation. Many of these mathematical understandings and processes are difficult and complex. Students gradually learn more and more about each idea over many years of schooling. Individual students will begin and end the unit with different levels of knowledge and skill, but all will gain greater knowledge of visualizing spaces, measuring volume, and using volume relationships.

In the *Investigations* curriculum, mathematical vocabulary is introduced naturally during the activities. We don't ask students to learn definitions of new terms; rather, they come to understand such words as *factor* or *area* or *symmetry* by hearing them used frequently in discussion as they investigate new concepts. This approach is compatible with current theories of second-language acquisition, which emphasize the use of new vocabulary in meaningful contexts while students are actively involved with objects, pictures, and physical movement.

Listed below are some key words used in this unit that will not be new to most English speakers at this age level, but may be unfamiliar to students with limited English proficiency. You will want to spend additional time working on these words with your students who are learning English. If your students are working with a second-language teacher, you might enlist your colleague's aid in familiarizing students with these words, before and during this unit. In the classroom, look for opportunities for students to hear and use these words. Activities you can use to present the words are given in the appendix, Vocabulary Support for Second-Language Learners (p. 109).

box, layer, stack   Students explore how many cubes a *box* (rectangular prism) will hold and come to see the cubes in a box as being *stacked* in *layers*.

method, predict   Students determine a *method* to *predict* how many cubes will fit in a box of given dimensions.

package   Students use the term *package* to signify more than one cube put together into a larger rectangular solid.

space   In the third investigation, students learn that the *space* in a rectangular solid (whether a paper box or a classroom) can be measured with cubic units.

model   In the final project of the unit, students work in small groups to build a *model* of their choice from various solids (cones, cylinders, prisms, pyramids) constructed from paper.

Investigations

The Packaging Factory

What Happens

Sessions 1 and 2: How Many Cubes? Student pairs look at pictures or written descriptions of rectangular boxes and predict how many unit cubes will fit inside the boxes. They check their predictions by building the boxes and filling them with cubes. Students then describe their method for making predictions and try to convince their classmates that their method will always work.

Sessions 3 and 4: Doubling and Halving Boxes of Cubes Student pairs determine the dimensions of boxes that will hold twice as many cubes as a box that is 2 by 3 by 5. They then determine the dimensions of boxes that will hold half as many cubes as a box that is 2 by 4 by 6.

Mathematical Emphasis

- Seeing 3-D rectangular arrays of cubes as congruent layers
- Determining the number of cubes that fit in a rectangular box
- Using multiplication to find the number of cubes in a box
- Determining the relationship between the number of cubes that fill a rectangular box and the dimensions of the box

What to Plan Ahead of Time

Materials

- Interlocking cubes: 70 per pair (All sessions)
- Scissors and tape for each pair (All sessions)
- Calculators: available (All sessions)
- Overhead projector (Session 1)
- Furniture catalogs that give the dimensions of items: 1 page per pair (Sessions 3–4)

Other Preparation

- If this is your first time through the unit, read the investigations with cubes, scissors and tape, and student sheets at hand. Do each problem, ideally with a partner; this is the best way to understand the activities and be prepared to help your students.

- Become familiar with the extension activities. Students generally complete geometry activities at different rates, and extensions are good for those who finish the activities early.

- If interlocking cubes are new to your class, make them available for unstructured experimentation before you start the unit.

Note on Cube and Grid Size The cubes that you use for this unit need to fit exactly on the grids used to make boxes. Check to see that your copy machine preserves the size of the grids in the blackline masters, and make all copies from original masters, rather than from a copy. If you are not using ¾-inch cubes, use graph paper to match the size of your cubes.

- Duplicate student sheets and teaching resources (located at the end of this unit) as follows:

For Sessions 1 and 2

Family letter (p. 112): 1 per student. Remember to sign it before copying.

Student Sheet 1, How Many Cubes? (pp. 113–114): 1 per pair, and transparency of first page

Student Sheet 2, A Method for Predicting (p. 115): 1 per student

Student Sheet 3, Predicting Numbers of Cubes (p. 116): 1 per student (homework)

Student Sheet 4, Painting Cubes (p. 117): 1 per student, as needed for extension

Packaging Factory (p. 118): 1 transparency

More Boxes for Predicting (p. 119): 1 copy to show to individual students

Large Box (p. 120): 1 transparency

Three-quarter-inch graph paper (p. 154): 4–7 sheets per pair (for use in Sessions 1–4)

- Use graph paper to cut out a pattern for a box that is 2 by 3 by 2. You will fold and tape the box during a demonstration. Also put together an array of interlocking cubes, 2 by 3 by 2, to fit into the box. (Sessions 1–2)

- Make a demonstration box that is 2 by 4 by 6. (Sessions 3–4)

- If you plan to provide folders in which students will save their work for the entire unit, prepare these for distribution during Session 1.

How Many Cubes?

What Happens

Student pairs look at pictures or written descriptions of rectangular boxes and predict how many unit cubes will fit inside the boxes. They check their predictions by building the boxes and filling them with cubes. Students then describe their method for making predictions and try to convince their classmates that their method will always work. Their work focuses on:

■ determining the number of cubes that fit in a box by examining pictorial and written descriptions of the box

■ developing, describing, and justifying a strategy for determining the number of cubes that fit in a box

Materials

■ Interlocking cubes (70 per pair)

■ Graph paper to match your cubes (3–5 sheets per pair)

■ Scissors and tape for each pair

■ Student Sheet 1 (1 per pair)

■ Student Sheet 2 (1 per student)

■ Student Sheet 3 (1 per student, homework)

■ Student Sheet 4 (1 per student, as needed for extension)

■ Family letter (1 per student)

■ Overhead projector

■ Transparencies of Student Sheet 1 (page 1), Packaging Factory, and Large Box

■ More Boxes for Predicting (1 copy)

Ten-Minute Math: Counting Around the Class Ten-Minute Math activities are intended for use outside your regular math time, in any spare 10 minutes during the class day. Once or twice during this investigation, spend some time on the activity Counting Around the Class, using the decimal variation. For suggestions about how to maximize all students' participation and confidence, see the full description of Counting Around the Class (p. 105).

Start with 0.5. Ask students to predict what number they'll land on if they count around the class exactly once by that number. Encourage students to talk about how they could figure this out without doing the actual counting.

Then start the count: The first student says 0.5 ("five tenths"), the next 1, the next 1.5 ("one and five tenths"), and so forth. Write the numbers on the board as students count so they can see the visual pattern.

Stop two or three times during the counting to ask questions like this:

We're at 7.5—how many students have counted so far?

After counting around the class once, compare the actual ending number with the students' predictions.

You might repeat the activity with students saying the fraction names for each decimal, in simplest terms (one-half, one, one and one-half, and so forth). Write the fractions above or below the decimals you wrote on the board as they counted.

As students gain more experience, try counting by 0.25, 1.5, 0.75, 0.2, and 0.3.

Packing Cubes in Shipping Boxes Display the Packaging Factory transparency on the overhead, covering the bottom half. Establish the scenario for the first two investigations in this unit by presenting the following information.

Let's suppose that you are working at a packaging factory. Your company makes cardboard boxes of different shapes and sizes, for packaging and shipping products.

One product you package is little ornaments. Each ornament is the same size, and is packaged in a little cube *[point out the cube on the transparency]*. **You often need to ship more than one ornament at a time, so you pack the cubes in larger shipping boxes** *[point out these]*. **Your first job for the factory is to solve this problem:**

> **How can you accurately predict how many of these cubes will fit in a shipping box of a particular shape and size?**

Once you have found a way to predict the number of cubes, you will need to explain your method in writing. That way, other people working at the factory can use your method, too, and know which box to make for an order of any size.

To clarify the task and introduce the box-pattern diagrams students will be seeing, uncover the bottom half of the transparency. Then show the box pattern you have made from graph paper, demonstrating how the pattern folds into a box, as shown on the transparency. Hold up a single interlocking cube, which students will use in this investigation to represent one ornament.

Place an array of interlocking cubes, 2 by 3 by 2, inside the box to show how it fits. Note that the cubes must be connected to fit inside the box properly, and the box should be taped so the sides meet but do not overlap. Keeping one side of the pattern untaped makes it easier to place the cubes into the box; point this out to students and suggest that they do the same when making their patterns.

This demonstration is *not* intended to give students any hint of how to solve the problems in this activity. Therefore, don't discuss the number of cubes in your box or any methods for figuring the number of cubes at this time.

How Many Cubes?

Clarifying the Terminology To avoid confusion, establish the difference between cubes, packages, and boxes. For these investigations:

- Single cubes are called *cubes*.
- Cubes put together into rectangular solids (such as the 2 by 3 by 2 array) are called *packages*.
- The rectangular boxes made from paper and filled with cubes or packages are called *boxes*.

Explain that the best way to talk clearly about three-dimensional objects is to explain our thinking with the relevant boxes, packages, and cubes in hand. This will help students keep track of the three terms.

Note: The boxes students will be working with in the first three investigations are, like most packing boxes, *rectangular prisms*. A rectangular prism has six faces or sides, all of which are rectangular regions. During Investigation 4, students will use the term *rectangular prism* when they compare different solids, but here, in keeping with the factory scenario, the everyday word *box* will be used instead.

Introducing the Activity Each pair will need 70 interlocking cubes, 3–5 sheets of graph paper to match the size of the cubes, scissors and tape, and a copy of Student Sheet 1, How Many Cubes?

Display a transparency of the first page of Student Sheet 1 with the overhead, covering all but the pattern and picture for box 1.

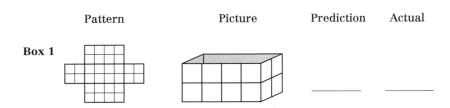

We're going to do some problems to help us find a way to predict the number of cubes that will fit in a box. Here is a picture of a shipping box, and a pattern that you could fold up to make the box. First, predict how many cubes fit in the pictured box. Write your prediction in the blank *[point]*.

Next, draw the box pattern on graph paper *[hold up a sheet],* cut it out, fold it, and tape the edges to make the box (leaving one edge untaped). Fill your box with cubes to check your prediction, then write down the actual number of cubes that fit in the box. Be sure you check your prediction for each box before going on to the next problem. That way, you'll get better at predicting as you go along.

Go ahead and start on box 1 now. Predict, build the box, and check the count with cubes. Then go on to the remaining boxes.

Making predictions before they use the cubes to find answers helps students form and organize mental images or models of the cube arrays. As students check their predictions, they refine these models and their ideas about figuring the number of cubes that will fit. Encourage students to think about discrepancies between their predicted answers and the actual count.

Later on, you're going to write about a way to predict the number of cubes that fit in any box. Be thinking about what you might write as you work on these problems.

Making Box Patterns At first, some students may not understand exactly how the patterns fold into boxes, how parts of a pattern correspond to parts of a box, or how to draw the pictured pattern on graph paper. Many will have difficulty making a pattern when one isn't shown (as for boxes 5 and 6). However, as students work with the patterns, most will eventually be able to design patterns themselves.

As you walk around the room and observe students working, ask questions to help those with difficulties:

Is the pattern you've drawn on the graph paper exactly like the one pictured? Does each part of your pattern [bottom, sides] match the same parts in the pictured pattern?

Would drawing the *bottom* of the box first help? Where do you think you should draw the bottom so that the sides of the box will also fit on the graph paper?

What part of the pattern will make *[point to a side on the picture of the box]* this side of the box?

Which side of the box does this flap on the pattern make?

While some students will correctly draw a pattern by tracing its perimeter—4 over, 2 down, 2 over, and so on—drawing the bottom and sides separately seems more reliable and helps students understand the structure of the box.

Encourage students who are successfully drawing a pattern to explain their strategy to students who need help.

Keeping the General Method in Mind Not all students will be able to think about formulating a generalized method for making predictions while working with these problems; they will focus only on finding answers for individual boxes. From time to time, remind students that their goal is to come up with a general method for predicting the number of cubes in a box *without* building it. Some students will begin to formulate a general method but have difficulty expressing it. As you circulate, ask students to tell you how they are predicting and finding the actual number of cubes in several specific boxes. This will help prepare them to write about a general method in the next activity.

Students will differ in how they approach and perform this task. Some will need to make every box to find correct answers; others will learn quickly to predict correctly from the pictures. Do not push for uniformity of method. Rather, encourage students to make sense of their answers, to improve their predictions, and to develop descriptions of their methods.

The **Teacher Note,** Strategies for Finding the Number of Cubes in 3-D Arrays (p. 26), will help you understand how students are progressing in their thinking; it also suggests ways to help students who are not making progress. Don't be alarmed if, by the end of the first session, many students are still using prediction strategies that give incorrect answers. Many will not discover accurate prediction strategies until the second session.

For Additional Help If there are students who are having difficulty, either with making boxes or with thinking in terms of layers of cubes, meet with them individually or in small groups. Use the resource sheet More Boxes for Predicting (p. 119) to offer some guidance. Each of the boxes on this sheet is one layer higher than the preceding box; the purpose of this sheet is to help students focus on the layers of cubes and see how they might determine the number of cubes that would fit in a box without counting them individually. Ask students to look for relationships between the boxes and to use these relationships to determine the number of cubes in successive boxes.

How are the first two boxes alike? How are they different? Can you find the number of cubes in the second box by using your answer for the first one?

If students don't see the pattern, do *not* point it out; instead give them cubes and graph paper to build the boxes as necessary. You can illustrate the same principle with actual cube packages. For example, put together four or five layers that are 3 by 2. Lay one on a table and ask students how many cubes are in this package. Lay another layer on top of the first and ask how many cubes are in the new package, and so on.

After students have completed Student Sheet 1, through box 6, pull the class together for a brief discussion. Point out that there may be many ways to make a good prediction—there is no single right way. Ask students to describe and show the different methods they used. Many students will use the same methods, so this can be a short discussion. Remind students that the best way to talk about the boxes and the cubes is to show *and* tell; otherwise it is almost impossible to communicate clearly.

How did you predict the number of cubes that fit in each box before you built it? Show us what you did. Were your predictions correct?

Will your method always work? How can you convince the class that your method will always work?

Did anybody try any prediction methods that didn't work? What were they? Why didn't they work?

Distribute and read aloud Student Sheet 2, A Method for Predicting. Be sure the students understand the tasks: describing a general method for finding the number of cubes and applying the method on a box with larger dimensions. Remind students to think about how they did the six problems on Student Sheet 1. Their method should work for any box, whether they start with a box pattern, a picture of the box, or only a written description of the box. Students may consult with each other, but they should write their own procedures. Encourage students to draw pictures to help explain their ideas if necessary.

❖ **Tip for the Linguistically Diverse Classroom** For students who are not writing comfortably in English, offer the alternative of either modeling their procedure or drawing a storyboard of the process.

Consider the following when you assess students' written work:

■ Does the method work? That is, does it result in correct predictions?
■ Are students using a layer approach or some less efficient strategy?

The **Teacher Note,** Strategies for Finding the Number of Cubes in 3-D Arrays (p. 26), describes the different kinds of strategies students may develop at different levels of understanding. To help students who are having difficulty devising a procedure that works, refer to the use of the sheet More Boxes for Predicting (p. 119) in the previous activity.

Some students may at first have difficulty describing their strategies in writing. If so, consider making this work part of a writing activity outside of math time. Students can trade papers with a partner and test the clarity of their writing by having their partner follow the method they described, using a box from Student Sheet 1. Then the writer modifies the description as needed.

If, when you are assessing students' strategies, you find it difficult to understand their written method, look back at their work on Student Sheet 1. Using their answers to these problems in conjunction with their written method should give you a pretty good sense of what the students are thinking. If you still can't follow some students' work, ask them to choose a problem from Student Sheet 1 and tell you how they did it.

In the second task on Student Sheet 2, it is impractical for students to actually build a box this large or use cubes to fill it. The problem encourages them to abstract the methods they used earlier. Also, they will have to convince their classmates that they are correct through argument, rather than by filling a box with cubes.

First thing we always did was find out how many cubes would fit in the bottom. The second thing we did was times that number by how many times it went up. That gave us our prediction.

procedure: first you count the bottom and then the rows on the sides. Then we multiply the bottom by the sides rows.

Count the squares across the bottom and down the bottom. Then times those numbers. Then count how many squares up and use the answer from the first thing and times it by how many squares up.

Example:

$3 \times 5 = 15$

$15 \times 3 = 45$

After everyone has finished their writing on Student Sheet 2, hold a class discussion of problem 2.

How did you figure out the number of cubes in problem 2? We don't have enough cubes to check the answer, so how do you know that you're right?

Arguments that students might use include these:

> On the other boxes, we found the number of cubes on the bottom, then multiplied by how many cubes high the box was.

> We multiplied the length by the width to find the number in the bottom layer. We then multiplied by the number of layers—that's the height.

Show the Large Box transparency during the class discussion of problem 2 (but not beforehand) so that students can show what they are talking about. You might ask:

How do we know that this is what the box in problem 2 looks like? How many cubes are in the bottom layer? How do you know? How many layers are there? How do you know? How many cubes will fit in the box? How do you know?

Sessions 1 and 2 Follow-Up

- Send home the family letter.
- Give each student a copy of Student Sheet 3, Predicting Numbers of Cubes. In the second problem, students are encouraged to extend their thinking to prisms with nonrectangular bases. Note that some students will not see the figure in the second problem in terms of layers, but instead as two or more rectangular prisms. Both approaches are valid and useful.

  After making their predictions about the number of cubes in each package, students may check their answers with cubes at the beginning of the next session.

 **Homework**

Painting Cubes For students who have the time, distribute Student Sheet 4, Painting Cubes. These problems are challenging and will help students further analyze how the cubes are organized in a 3-D array.

 **Extension**

Answer key: **1.** 60 cubes **2.** 6 cubes **3.** 22 cubes **4.** 24 cubes **5.** 8 cubes **6.** no cubes **7.** no cubes **8.** no cubes

Strategies for Finding the Number of Cubes in 3-D Arrays

For students to determine how many cubes are in a 3-D array, they must mentally construct an image or model of the set of cubes. We have observed a variety of ways that students do this. Here are some of these ways, listed in increasing order of the students' ability to see the whole and its parts in an organized manner. You will see your students progress from less organized to more organized concepts as the unit progresses. But the progress may be slow, and the same students may approach different tasks with different levels of understanding.

Seeing Arrays as Unstructured Sets   Whether given an actual cube package, a picture of a cube package, or the box that contains that package, students act as if they see no organization of the cubes. In this case, students usually count cubes one by one, and almost always lose track of their count. For these students, the task is like counting a large number of randomly arranged objects.

 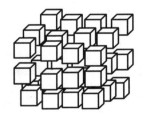

Seeing Arrays in Terms of Sides or Faces
Many students approach 3-D arrays of cubes by thinking only about the sides of the rectangular prism formed by the cubes. These students might count all or some of the cube faces (squares) that appear on the six sides. With this method, edge cubes are often counted more than once, and cubes in the middle are missed.

 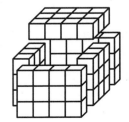

Thus, a box of 36 cubes, 3 by 4 by 3, might be counted as 54 cubes—the front, back, and top each as 12, the right and left sides each as 9.

Most students with this "sides" conceptualization use it consistently, whether they are looking at pictures of boxes, box patterns, or the actual cube configurations. For example, after one student put together the package that fit into box 1 on Student Sheet 1, she counted 8 faces on the front and 4 on the right side, thus counting the 2 cubes in the column shared by these sides twice.

She then doubled this amount to account for back and left sides, added 8 for the top, and got a total of 32.

Students who see cube arrays in terms of their faces do not necessarily think of arrays as hollow; they simply think that their method enumerates all the cubes inside and out.

Students who think of 3-D arrays in terms of their sides often benefit from working on more problems in which they must think about the correspondence between squares on the box patterns and cubes that fit into the box. For example, cut out the pattern for a box that is 2 by 2 by 2, but do not tape it. Put together a package of cubes, also 2 by 2 by 2, and place it on the base of the pattern. Ask how many cubes are in the package. If a student says that there are 4 cubes for each of the 4 squares on the side flaps (getting 16), ask him or her to count the cubes in the package by taking it apart.

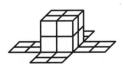

Then ask the student to reassemble the cube package and place it back onto the pattern.

Continued on next page

Do you see how the squares on the sides of the pattern match up to the cubes in the package? Why did counting all the squares on the sides not work?

Some students will need even more specific questions on the correspondence between squares in the pattern and cubes. For example, while pointing to a square on a pattern, you might ask what cube on the package it matches.

Seeing Arrays as Having Outside and Inside Parts Students who take this approach try to count both the outside and inside of the 3-D array, sometimes doing it correctly, but more often incorrectly.

■ **Correct Counting** One student counted the cubes visible on the front face (12), then counted those on the right side that had not already been counted (6). She then pointed to the remaining cubes on the top, and for each, counted cubes in columns of three: 1, 2, 3; 4, 5, 6; … 16, 17, 18. She then added 12, 6, and 18.

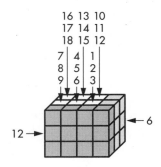

■ **Incorrect Counting** One student counted all the outside cube faces of this same array, getting 66. He then said that there were two cubes in the middle, arriving at a total of 68.

Seeing Arrays in Terms of Rows or Columns Students count the cubes in successive rows or columns by ones or by skip counting. In the strategy diagrammed below, the student counted three cubes for each of the six columns.

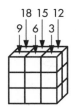

Seeing Arrays in Terms of Layers Students determine the number of cubes in one layer, then multiply or use repeated addition to account for all the layers. The layers can be vertical or horizontal, and students often use one of the visible faces in a picture as a representation of a layer. Other students look at a box pattern, see the bottom as representing a layer, then determine the number of layers by looking at the sides of the pattern. ("It goes up 3.") Many students who use layering still need to count the cubes in a layer one by one.

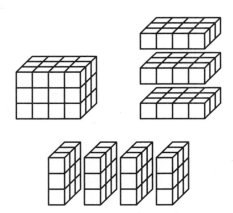

Continued on next page

Seeing Arrays as Layers Described by Dimensions Some students understand how dimensions can be used to describe and count the cubes in an array.

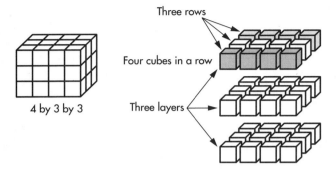

4 by 3 by 3

Three rows

Four cubes in a row

Three layers

Students might reason that the length gives the number of cubes in a row and the width gives the number of rows in a layer, so the number of cubes in a layer is the product of the length and width. Because the height gives the number of layers, they multiply the number of cubes in a layer by the height to find the total number of cubes in the array. (Not all students refer to the dimensions as length, width, and height.)

To help students who think of arrays in terms of layers, but who cannot use dimensions to find the number of cubes in these layers, give them more problems similar to problem 6 on Student Sheet 1, with only written descriptions of the box dimensions. Keep the dimensions of the boxes small so that students can check their answers by making boxes and filling them with cubes.

The Learning Process Students gradually progress to more powerful ways of conceptualizing cube configurations as they have repeated experiences with making predictions, building boxes, filling them with cubes, enumerating the cubes, and discussing their ideas with classmates.

Avoid showing students a standard method for finding the number of cubes; let them generate their own strategies. Showing them your methods leads students to follow those methods and can discourage them from generating their own personally meaningful strategies.

If a student uses the standard formula or algorithm for volume, it is essential that he or she be able to explain *when* it applies and *why* it works (for example, by explaining how many layers there are and how many cubes per layer). Be aware that some students will use the algorithm without explicitly using the terms *length, width,* and *height* or the symbols *l, w,* and *h*.

Expected Progress Do not expect all students to be able to think in terms of layers by the end of Investigation 1. It is sufficient if they have developed a prediction procedure that gives correct answers. Despite your efforts to promote it, some students may not arrange cubes into layers until after Investigation 2. There is no need to delay Investigation 2 until all students start layering. Also be cautious about pushing students to use dimensions. Too much emphasis on this can pressure students to adopt a numerical procedure (multiplying the dimensions) that they do not understand, causing them to abandon a visual method that makes sense to them. Ultimately, you can expect most students to adopt a layer approach for finding the number of cubes that fit in a box, and many will correctly use dimensions.

DIALOGUE BOX

Understanding Multiplication and Arrays

Fully understanding why multiplication works to determine the number of objects in a rectangular array puzzles many students. Julie regularly determined the number of cubes in 3-D arrays using a layer strategy. She also had discovered that the number of cubes could be found by multiplying the three dimensions. But as her class discussed these strategies, Julie was puzzled. And even though almost every student in the class had discovered, and was routinely employing, a layer approach, not one of them had an immediate answer to Julie's confusion.

Julie: The corner cube gets counted once when you find the length, once for the width, and once for the height. So the answers we're getting should be wrong. But I think they're right.

[To simplify the problem, the teacher shows a one-layer array of cubes, 4 by 3.] **How many cubes do you see here?**

Corey: There's 12; 4 times 3.

Julie: I know the answer is 12. But when you multiply, you count the corner cube once for the length and once for the width. So you count it twice.

Amir: That cube is in both the width and the length. It's OK to count it twice.

[The teacher separates the array of cubes into 3 rows of 4, and points to the cubes in one row.] **1, 2, 3, 4. What am I counting here?**

Leon: Cubes.

[Pointing to the three rows] **1, 2, 3. What am I counting here?**

Julie *[excitedly]*: Rows of cubes. You're not counting cubes this time. So first you count cubes, then you count rows.

Leon: So you're not really counting the cube twice. We got it!

Julie's question posed a real conundrum for the students. They were sure that multiplying the length times the width gave the number of cubes in a rectangular array. Almost all of the students justified this procedure by saying that they were multiplying the number of cubes in a row by the number of rows, thus satisfying the traditional criterion that they understood the procedure. But initially, they did not recognize what they were counting; they were applying multiplication mechanically.

Doubling and Halving Boxes of Cubes

What Happens

Student pairs determine the dimensions of boxes that will hold twice as many cubes as a box that is 2 by 3 by 5. They then determine the dimensions of boxes that will hold half as many cubes as a box that is 2 by 4 by 6. Student work focuses on:

- determining the number of cubes that fill a rectangular box of given dimensions
- exploring the relationship between the numbers of cubes that fill a rectangular box before and after its dimensions are changed

Materials

- Completed Student Sheet 3 (from home-work)
- Catalog pages showing furniture dimensions (1 per pair)
- Completed Student Sheet 1 for reference
- Interlocking cubes (70 per pair)
- Graph paper to match your cubes (1–2 sheets per pair)
- Tape and scissors for each pair
- Demonstration box (2 by 4 by 6)

Activity

Checking Homework Predictions

Students check the predictions they made on Student Sheet 3 (homework) by making the pictured packages with cubes. Below their predictions, they record the number of cubes it took to make each package. They can do this early in the school day or at the beginning of math class.

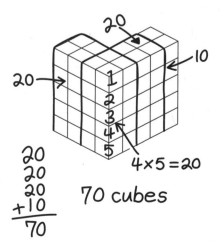

How We Give Dimensions

In your job at the packaging factory, you've been making boxes of different sizes. If you were going to sell these boxes, how would you describe the different sizes? Let's look at some pages from a catalog and find out how another business describes the size of objects.

Give each pair of students a page from a catalog that advertises furniture—desks, files, bookcases, sofas, dressers, and the like. Ask students to find some description of the size of an object in the catalog items. Write some of the examples they find on the board. For example:

> Desk 30"h x 59"w x 30"d
>
> 6-shelf bookcase 72"h x 12"d x 36"w
>
> Bed 70" x 48"

How are the sizes of things described in the catalog? The numbers used to describe the size of an object are the *dimensions* of the object. What does the *x* mean in these descriptions?

Be sure the discussion clarifies that we read the *x* in these expressions as *by* rather than *times*.

When we see dimensions written like this *[write 3 x 4 x 5]*, **with an *x* between the numbers, we read this as "3 by 4 by 5."**

Keep in mind that the *x* notation in dimensions can be confusing to students because they are accustomed to reading *x* as *times*. Ask students how they know what the three measurements for each item refer to.

How do we know what these numbers mean?

Once students recognize that each measurement describes a particular dimension of an object, ask if the order in which the measurements are given is important.

Do the examples you found in the catalog advertisements always give the numbers in the same order?

If you have a typical collection of catalogs, some items will have unlabeled dimensions or missing dimensions, and dimensions will be given in various orders. From these examples, students should see that order is not significant. For a further example, refer students to their completed Student Sheet 1 and draw attention to box 1.

How would you describe this box, using three numbers?

Help students see that while the numbers themselves are important, the order in which they are listed is not. For this activity, students need to agree that the order of the dimensions is irrelevant in describing the shape of the box. For example, we could describe box 1 as 4 by 2 by 2, or 2 by 4 by 2, or 2 by 2 by 4. Many students will believe that the order of the numbers depends on the orientation of the box. Also, many do not like the ambiguity of omitting labels, so they will very clearly describe the box as "4 by 2 on the bottom, and 2 high." The **Dialogue Box,** Discussing Dimensions (p. 36), illustrates what you might expect to happen during this discussion.

Doubling the Number of Cubes

Let's say we have a box that is 2 by 3 by 5 *[write these dimensions on the board].* **The factory wants us to build a box that will hold twice as many cubes. What are the dimensions of a box that contains twice as many cubes as a box that is 2 by 3 by 5? Are there other dimensions that will also work? See how many boxes you can find that will hold twice as many cubes.**

Working in pairs, students try to find several boxes that hold twice as many cubes as the original box. They then check their answers by making paper boxes and filling them with cubes. Distribute graph paper as they are ready for it. The **Dialogue Box,** Common Student Strategies for Doubling (p. 37), illustrates the widely varying understandings that students might apply to this problem.

The goal for this activity is for students to see the connection between changing the dimensions of a box and changing the number of cubes that fit in the box. Students need to think about the relationship between an object (a package of cubes) and the numbers that describe that object (the dimensions of the package).

As students solve this problem, they describe the proposed boxes in terms of dimensions. For example, they might describe a box in one of the following ways:

It's 4 by 3 on the bottom, and 5 layers high.

The box is 4 by 3 by 5.

It doesn't matter how students describe the dimensions, as long as their descriptions are easily understood by others.

Discussing Students' Solutions   After each pair has found two or three solutions to the doubling problem, bring the class together for a brief discussion. If students get confused describing the dimensions of a box, they might set it on a flat surface and walk around it to describe its dimensions.

What are the dimensions for boxes you found that will hold twice as many cubes? *[List these where everyone can see them.]* **What do you notice about these dimensions?**

One way to double the number of cubes is to double *one* of the original dimensions, but there are other ways. Some students will find possibilities such as 1 by 1 by 60, or 2 by 1 by 30. Although these dimensions are not connected to the shape of the original box, you still need to listen to students explaining why these boxes hold twice as many cubes as the original, and how they came up with their ideas. Many will explain that they simply looked for sets of three numbers that multiplied together to give 60.

How did you find the dimensions of boxes that contain twice as many cubes as the original box? How many of you started the problem by doubling all the dimensions? Why didn't that work?

Refer to the last extension activity (p. 35) for an interesting question about doubling all the dimensions.

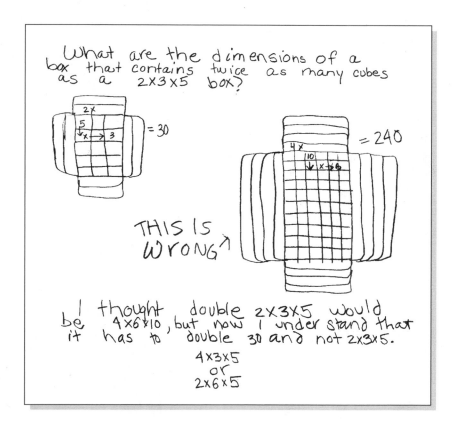

Halving the Number of Cubes

Hold up your demonstration box (2 by 4 by 6).

What are the dimensions of this box? I want you to find the dimensions of a box that holds half as many cubes as this box. Make a prediction, then check yourself. Is there more than one set of dimensions that would work?

If students have difficulty with the concept "half as many," ask:

If I have 10 cubes, what is *half as many* cubes? If I have 8 cubes, what is *half as many*?

On a sheet of notebook paper, students write their predictions, a description of the methods they used to check their predictions, and their final answers. They check their predictions using whatever methods they choose, including words and computations, drawing pictures, or building paper boxes.

For students who have difficulty getting started, suggest that they make and analyze their own paper box or package of cubes with the original dimensions, 2 by 4 by 6.

Look for the following in students' work:

■ Did they find a set of dimensions that works? There are many, but most students will find only one solution, and this is satisfactory.

■ Was their prediction correct? If so, this suggests that they have generalized the idea from their work with doubling. Of course, some students will make the same conceptual error they made on the doubling problem—they will take half of each of the original dimensions.

■ Did they find more than one set of dimensions that works? Students who have fully abstracted the idea will easily find several solutions.

```
1 x 4 x 6          In the beginning I
2 x 2 x 6          did a box of cubes
2 x 4 x 3          2x4x6 and tryed to figure
1 x 3 x 8          out how many ways there
1 x 2 x 12         were to put it apart into
                   2 boxes the same. Then
                   after word instead of
                   takeing them apart I did
                   the rest of the demensions
                   in my mind.
```

Sessions 3 and 4 Follow-Up

■ **Three Times the Number of Cubes** Challenge students to find a box that holds *triple* the number of cubes. Specifically, ask them to find the dimensions of a box that contains three times as many cubes as a box that is 2 by 3 by 5. What other dimensions would work?

■ **Making a Twice-Size Box Display** Challenge student pairs to find all the different sets of dimensions that describe boxes that will hold exactly double the number of cubes as a box that is 2 by 3 by 5. Together they plan and make boxes that hold twice as many cubes. They label the dimensions of each box and paste their boxes onto a large piece of paper. This makes a powerful visual display that illustrates how the shapes of these doubled boxes are related to the original.

Note: For any set of dimensions, there will be several different boxes that have those dimensions because the open top could appear in different locations. However, all of these boxes will have the same basic shape, and should be considered the same for this activity. Be sure to explain this to students working on this extension.

■ **Doubling All the Dimensions of a Box** Students who tried doubling all the dimensions of a box to double the number of cubes it would hold found that this approach made a box that was too big. You might challenge them to explore this question:

Suppose I make a new box by doubling all the dimensions of my original box, which is 2 by 3 by 5. How does the number of cubes in the new box relate to the number in the box I started with?

You may hear the correct answer expressed in unique ways:

It eighted [as compared to doubled].

It quadruple-doubled.

D I A L O G U E B O X

Discussing Dimensions

This class is exploring How We Give Dimensions (p. 31), looking at the pages of catalogs that offer furniture for sale.

When we see numbers describing the shapes and sizes of the objects in the ads, how do we know what they are talking about?

Robby: We knew because they told us what the numbers mean: 29 *wide* by 80 *long* by 22 *high*.

What other descriptions did you find?

Heather: 39 by 80 by 72.

Marcus: 48 wide, 28¼ deep, 50 inches high.

Katrina: 36 wide times 25 deep times 60 inches high.

Does it say "times" in the ad?

Katrina: No, but they're *x*'s.

What do the rest of you think the *x*'s mean? Katrina thinks they mean multiplication.

Greg: Maybe it means that the item is adjustable.

Maricel: I think they're abbreviations for "by." [*Most students agree.*]

Do the catalogs always give the same order for the numbers? Suppose we are looking at a dresser for a room. Does it matter which measurements—length, width, height—are given first?

Sofia: If they tell what the measurements are, it doesn't matter, the order. But if they don't tell you, like in Heather's, then it could be confusing.

Heather: Yeah, you might need it 39 deep, but you don't know which number is for deep.

OK, let's go back and try to describe the very first box we made [box 1 on Student Sheet 1].

[The teacher holds up a paper box that is 4 by 2 by 2.]

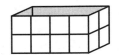

Matt: It's 4 by 2 by 2.

Duc: 4 long, 2 high, and 2 wide.

Katrina: 4 long, 2 high, 2 deep.

Heather: 2 by 4 by 2.

Would you all agree then, that 4, 2, and 2 are the numbers that describe the box? Does the order of the numbers change the shape of the box?

Amy Lynn: It doesn't matter how you describe it, it still holds the same amount.

Katrina: If you change the way you hold it, you describe it differently, but it's the same shape.

[The teacher now turns the box to various orientations, asking the same questions each time.] **Is this box 4 by 2 by 2?**

After each orientation, the students answer yes, and come to see that the order of the numbers doesn't really matter.

Common Student Strategies for Doubling

Students are working in pairs on the Doubling the Number of Cubes activity (p. 32). Noah and Lindsay have started by making a box that is 2 by 3 by 5 and filling it with a package of cubes with the same dimensions. To make a box that holds double the number of cubes, Noah thinks they should double each dimension, making a package 4 by 6 by 10. He looks at the original package and says they need to add onto it by layers, so he and Lindsay add 5 more layers. Noah turns the package onto its side so it's 3 cubes high. He says that now they have to build this side up 3 more, so it's 6 high. The teacher has observed what just happened.

How could you figure out how many cubes are in the original cube package by looking at the dimensions?

Noah: Multiply 2 times 3 times 5, so 2 times 3 = 6 and 6 times 5 = 30.

What would twice that many be?

Lindsay: 30 times 2 equals 60. Our new box should hold 60 cubes.

How many cubes will a box that is 4 by 6 by 10 hold?

Noah: We can multiply $4 \times 6 \times 10$ to find the answer. *[He uses a calculator.]* $4 \times 6 = 24$ and $24 \times 10 = 240$.

Lindsay: That's too many. So it's not doubled. I think we did something wrong.

Maybe it would help if you would go back and look at the first box you built. Try to use it to help you think about a new box that will hold twice as many cubes.

[The teacher later discovers other student strategies in the whole-class discussion of students' solutions.]

How did you find the dimensions of a box that contains twice as many cubes as a box that's 2 by 3 by 5?

Kevin: We kept drawing patterns until we found one that held 60 cubes. I know it's double because I filled it with 60 cubes.

Shakita: We made two packages out of cubes that were 2 by 3 by 5. When we put them next to each other one way, we got a package that's 4 by 3 by 5. When we put them next to each other another way, we got a 2 by 6 by 5, then we got a 2 by 3 by 10. *[She demonstrates the three different positions for their two packages.]*

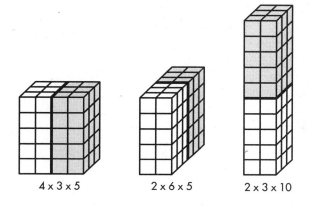

4 x 3 x 5   2 x 6 x 5   2 x 3 x 10

Noah and Lindsay's idea of doubling all the dimensions is common; they are thinking about the numbers but not checking to see if their numerical reasoning makes sense spatially.

Kevin's group explored patterns until they found a box with the desired number of cubes. They didn't see how the dimensions of the original box were related to those of the doubled box.

Shakita's group found a concrete way to generate several spatially meaningful solutions to the problem.

All these students needed opportunities to test their ideas with cubes and sometimes the calculator. Additionally, for Noah and Lindsay, the teacher played a critical role in redirecting their attention to focus on the total number of cubes.

Packing Problems

What Happens

Sessions 1 and 2: Packing Packages of Different Sizes   Student pairs find ways to accurately predict, then determine how many packages of different sizes will fit in a given box. They check their work by making the box and filling it with packages of cubes.

Sessions 3 and 4: Designing Boxes   Students design a single box that can be completely filled with each of four or five different-shaped rectangular packages. They pack the box with only one type of package at a time, and it must fill the box to the top with no gaps.

Session 5: More Packing Problems   Students predict how many two-cube packages fit into boxes that are marked off by rectangles, not squares. This activity uncovers a misconception many students have about multiplication and arrays.

Mathematical Emphasis

■ Organizing rectangular packages so that they fill rectangular boxes

■ Developing strategies for determining how many rectangular packages fill a box

■ Designing boxes to hold packages of different sizes

■ Understanding the relationship between the dimensions of a box and how many rectangular packages fill the box

What to Plan Ahead of Time

Materials

- Interlocking cubes: 70 per pair (All sessions)
- Scissors and tape for each pair (All sessions)
- Overhead projector (Sessions 1–2)
- Calculators: available

Reminder on Cube and Grid Size The cubes that you use for this unit need to fit exactly on the grids used to make boxes. If you are not using ¾-inch cubes, you will need to adjust the box patterns with Student Sheet 5 (pp. 122 and 124) and Student Sheet 7 (pp. 127 and 129–130), which are based on a ¾-inch grid. For example, if you use 2-cm cubes, you will need to enlarge these sheets by about 5% on a copier. Experiment with one enlarged copy first, and test it to be sure your cubes and two-cube packages match the grids.

Other Preparation

- To introduce the problems on Student Sheet 5, make packages A through E (see p. 121) from interlocking cubes. You might label the packages by writing the letters A through E on small stickers. Make a demonstration copy of box 1 from the pattern (p. 122). (Sessions 1–2)
- Duplicate student sheets and teaching resources (located at the end of this unit) as follows:

For Sessions 1 and 2

Student Sheet 5, How Many Packages? (pp. 121–124): 1 per pair, and transparency of first page

For Sessions 3 and 4

Student Sheet 6, Design a Box (p. 125): 1 per pair

Graph paper that matches your cubes: a small supply for use as needed

For Session 5

Student Sheet 7, More Packing Problems (pp. 126–130): 1 per pair

Student Sheet 8, Saving Cardboard (p. 131): 1 per student (optional, Extension)

Packing Packages of Different Sizes

Materials

- Prepared packages A through E and box 1, for demonstration
- Student Sheet 5 (1 per pair, plus transparency of first page)
- Interlocking cubes (70 per pair)
- Scissors and tape for each pair
- Overhead projector

What Happens

Student pairs find ways to accurately predict, then determine how many packages of different sizes will fit in a given box. They check their work by making the box and filling it with packages of cubes. Student work focuses on:

- organizing rectangular packages to fit in rectangular boxes
- developing strategies for determining how many rectangular packages will fit in a box

Activity

How Many Packages?

In this second investigation, students consider situations in which paper boxes are filled with rectangular packages that are not cubic in shape. Each package is made with several cubes. To avoid confusion, it is important to maintain the distinction between *cubes, packages,* and *boxes,* as explained in Investigation 1 (p. 20). Remind students that as they talk about their work in this unit, they always need to explain their thinking with the boxes, packages, and cubes in hand. Help students do this by setting a clear example.

Distribute to each pair the four-page Student Sheet 5, How Many Packages? If possible, also use an overhead to project a transparency of the first page of the student sheet for everyone to see.

In order to show the bottom of the boxes, the diagrams of boxes 1 and 2 on the student sheet are more complex than earlier diagrams. Having a model of box 1 can clarify the diagram for students. As you introduce the problems on Student Sheet 5, show students your actual cube packages A through E and the demonstration box 1.

Here we are back on the job at the packaging factory. We need to make boxes to ship different quantities of these five packages. *[Display the packages you have made from cubes and show how they correspond to the pictured packages A, B, C, D, and E.]* **Your job is to find a way to accurately predict how many of each package will fit in a box before it is made. You will be working with a partner on this job.**

You'll start with package A. How many of these do you predict will fit in box 1? *[Show your demonstration package A made with cubes, and demonstration box 1.]* **You'll write your prediction down** *[show where on the transparency of Student Sheet 5],* **then you'll check your prediction. To check, make your own box 1. The pattern for this box is on the second page of the handout. Use cubes to put together packages the size of package A into the box. See how many of these packages will fit in box 1, and write that down as the actual number** *[show where].*

When you have finished with package A, move on to package B. Make your prediction by looking at the picture or your paper box. After you've written down your prediction, check it with packages you make from cubes.

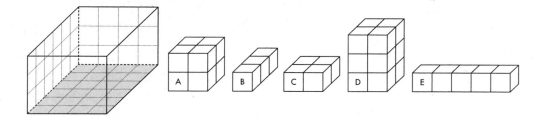

When students have completed the five problems for box 1, they proceed to box 2. They will make box 2 by adding strips to the sides of box 1, as described on the fourth page of Student Sheet 5.

As students work, circulate to check on their progress. By actually building and filling the boxes, students will recognize errors in their prediction strategies. If they don't predict accurately at first, encourage them to reflect on their work and develop a better method of prediction. For example:

You predicted 9 for package A, but only 6 actually fit. Why do you think your prediction was inaccurate? Is each package going to fill the box all the way to the top? Will some packages fit better than others? How can you tell?

In checking their answers, some students will need to completely fill the boxes with the packages. Others might make only one package and move it around the bottom of the box to determine the number in a layer. See the **Teacher Note,** Strategies for Finding How Many Packages (p. 43), for a description of the different strategies you might expect from students.

These problems will be difficult for many students. Most will have discovered a correct procedure for finding the number of unit cubes in a box in Investigation 1, but some of them will not understand why their procedure works. They will not see how the organization of cubes within the box (say, in terms of layers) corresponds to the numerical procedure they use to find the number of cubes.

Teacher Checkpoint

How We Counted Packages

To get a quick feel for how students are doing on these problems, collect their work on Student Sheet 5 to compare their predictions and actual answers. Students whose predicted and actual numbers are identical probably have discovered a valid strategy for determining the number of packages that fit in the boxes. You can get an even better understanding of students' thinking as you listen to them explain their strategies in small groups and the subsequent class discussion.

Class Discussion   After everyone has completed Student Sheet 5, hold a brief class discussion in which students demonstrate their strategies. For each box, select two pairs of students who differ in their answers or strategies and ask them to demonstrate their methods with the materials they used.

Students may have built with cubes to model the situation, drawn on graph paper, visualized, multiplied, and so on. No matter what strategy they use, each pair should try to describe and show it. Additionally, they should try to convince their classmates their answer is correct because their strategy is valid.

In the process of demonstrating, some may discover that their answers are not correct. When this happens, reassure students that mistakes are often part of the process of finding a good solution. Help them see how analyzing their errors can help them understand the ideas involved. Encourage students to view the classroom as a nonthreatening environment in which everyone tries to make sense of and question everyone else's ideas. Students need to be supportive of their classmates' mistakes and help each other understand why those mistakes happened.

Sessions 1 and 2 Follow-Up

 **Extension**

More Packing Problems   How many cube packages that are 2 by 2 by 2 will fit in a box that is 8 by 8 cubes on the bottom and 4 cubes high? How many packages that are 4 by 4 by 4 will fit in this same box?

Strategies for Finding How Many Packages → Teacher Note

To correctly predict how many packages fit in a box, students must accurately visualize the spatial organization of the packages within the box. At first, students may make a variety of mistakes. Through their use of cubes and paper boxes, or by explaining their strategies, they will discover their mistakes and why they occurred.

Two useful strategies we have seen students use are (1) visualizing where packages fit in a box, often thinking about layers, and (2) finding relationships between different packages. Following are examples of these two strategies and the most common mistakes students make in using them.

Visualizing Where Packages Fit in a Box For their work on Student Sheet 5, these three students accurately visualize the organization of packages A and B within box 1.

Tai [drawing circles around groups of 4 squares on the bottom of the box picture]:

4 here, 4 here,
4 here, 4 here,
4 here, 4 here.
There's 6. I can't
get any in the top
row [layer].

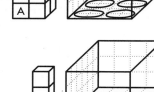

Becky [counting squares on the bottom]: So 24 go up.

Desiree: There's 8 on the bottom and 3 layers, so 24.

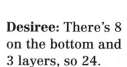

However, the next four students make some common mistakes in visualizing. The first error is losing track of one of the dimensions.

Kevin: I can fit 15 package E's—5 on the bottom *[motioning as indicated in the diagram]*, times 3.

Actually, only 4 packages fit on the bottom. Package E will *not* fit horizontally along the front.

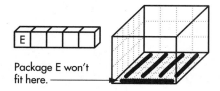

Package E won't fit here.

Another common error is losing track of the original unit.

Zach: With package E, there's 20 on the bottom. So 20 times 3 = 60.

Although he properly saw four package E's lying on the bottom, he counted each cube in the packages. Zach has forgotten that the original unit was a package of 5 cubes.

Students sometimes fail to see that the same package can be oriented different ways in the box.

Noah: We looked at the box picture [box 2] and visualized putting package E down in rows on the bottom and up 5 layers. That gave us 20. None fit over here because there's just 4 [squares] across.

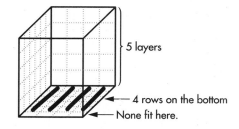

5 layers

4 rows on the bottom
None fit here.

Noah is visualizing only a horizontal package E. After actually putting five layers of four package E's into box 2, each laid horizontally, Noah's partner Lindsay sees their mistake. She notices that four more package E's will fit in the remaining space, if they are placed vertically.

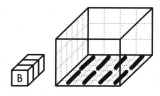

Continued on next page

Another common error is losing track of where packages are placed, so packages are double counted or missed.

Toshi: That's 6 on the bottom; then on top there are 2 in the back, 3 on the left side, and 3 on the right side. That's … 14 packages.

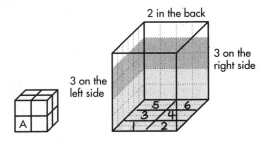

As Toshi and Cara look at box 2, they accurately visualize the bottom layer of 6 package A's. But on the second layer, they see two rows of 3 for the sides and 1 row of 2 for the back. They are double-counting the two in the back.

Finding Relationships Between Different Packages Comparing different packages directly (their shape, as well as the number of cubes they are made of) is an excellent strategy. For example, after finding that 6 package D's fit in box 1, Mei-Ling used this knowledge in thinking about how many package C's fit in the box.

Mei-Ling: I think 18 C's, because it's like D. D is 3 times larger [than C], so it'd be 3 times 6 [for C].

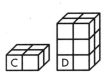

A common mistake students make when trying to relate different packages is ignoring package organization. That is, students do not think about how the packages actually fit in the boxes. For example, some students reason that the number of package B's that fit in a box is one-third the number of unit cubes. Or, as this student reasons:

Danny: There are 72 cubes in box 1 and package A is made up of 8 cubes, so 8 goes into 72 nine times. We can get 9 package A's in this box.

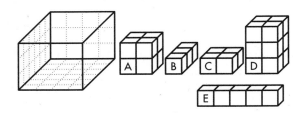

This strategy works for packages B, C, and D with box 1, but it doesn't work for A or E because copies of these packages do not completely fill the box. This strategy would work if we were allowed to break the packages apart, but we are not.

Designing Boxes

What Happens

Students design a single box that can be completely filled with each of four or five different-shaped rectangular packages. They pack the box with only one type of package at a time, and it must fill the box to the top with no gaps. Student work focuses on:

- determining how packages must be organized to fill a given box
- designing a box that can be completely filled with several differently shaped rectangular packages

 Ten-Minute Math: Counting Around the Class Once or twice during this investigation and the next, continue the activity of counting around the class by decimal numbers. If students have already counted by 0.5 and 0.25, try 0.75 and 1.5.

Materials

- Student Sheet 6 (1 per pair)
- Interlocking cubes (70 per pair)
- Scissors and tape for each pair
- Graph paper available as needed

Activity

Distribute a copy of Student Sheet 6, Design a Box, to each pair. If any students seem uncomfortable with the terms *cubes, packages,* and *boxes,* review examples of each again (see p. 20).

The packaging factory you work for would like to have one box they could use to ship many different packages. The box needs to be a size and shape that can be completely filled by any of the packages A, B, C, or D. Each kind of package must fill the box to the top, without leaving any gaps. Your job is to design and build the box, then report the dimensions to your boss.

Can you find more than one box that works? For each box that you design, record its dimensions and the number of each package it takes to fill it.

If you figure out a box that can be used for packages A through D, then try to figure out a box that can also be used for package E. How many boxes can you find that work for all five packages?

Design a Box

Working in pairs, students use graph paper, interlocking cubes, and whatever else they need to develop a design that will work for all the packages. To solve the problem, students will need to think about how the packages must be spatially organized to completely fill a box. The **Teacher Note,** Strategies for Designing Boxes (p. 48), explains some of the strategies you may observe in your class.

You will find that some students use multiples to solve these problems. For example, for package A to fill a box, all the dimensions of the box must be multiples of 2. Packages B and D require that one dimension also be a multiple of 3. Package E requires that one dimension be a multiple of 5. Recognize that this is an excellent strategy if students come up with it on their own, but avoid suggesting it yourself.

Note: Because this problem requires students to create a box that satisfies a number of constraints, they need plenty of time to make and test conjectures about possible solutions. Also, because students are creating something of their own, they become quite engaged in this activity. For these reasons, students usually need or want to spent two days on this task. Those who finish early can start the homework in class.

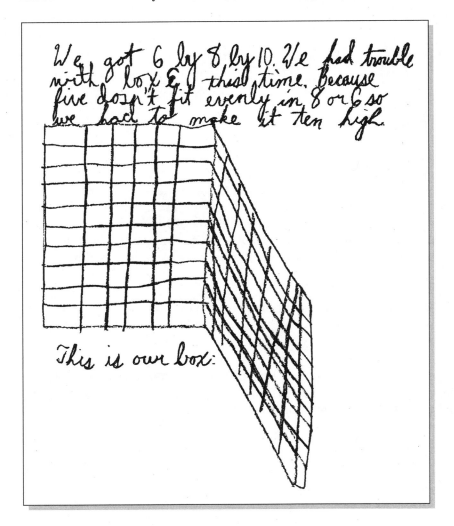

We got 6 by 8 by 10. We had trouble with box E this time. Because five doesn't fit evenly in 8 or 6 so we had to make it ten high.

This is our box:

Sessions 3 and 4 Follow-Up

Using their completed Student Sheet 5 for reference, students find as many ways as they can to pack box 1 with packages so that it is completely filled. This time, they can use *any combination* of boxes A–E that they choose. For example, a combination of 6 A's and 6 C's will completely fill the box. Students record all their solutions on notebook paper.

For their work on Student Sheet 6, Design a Box, students use a variety of strategies to generate possible dimensions for boxes that will hold packages A–D. Most students seem to need some type of concrete material to generate or test their ideas—cubes to make the packages, or graph paper to make box patterns.

Using Packages Many students generate ideas by stacking packages. Robby and Antonio found that 6 worked as a dimension because they could stack copies of any of the packages A–D and get a package that was 6 cubes high.

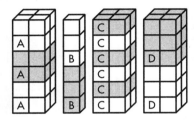

Alani and Rachel also used packages to generate ideas. But they put together *different* packages in various combinations to help them think about the problem. For instance, they put together one A, one C, one D, and two B's in the configuration shown here.

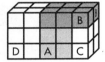

Alani tried to see if Package A would fill a box shaped like this by placing Package A in various positions alongside the configuration.

Alani: This won't work. It won't fit; there's extra space.

The girls then made a new configuration, again testing it by placing A's alongside.

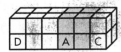

Alani: Three A's will fit. So will two D's.

Rachel: Four B's here and four B's here; that's 8. And six C's. We got it!

Using Patterns Greg and Duc think that a box 6 by 6 by 6 will work for packages A–D. They draw a pattern for a such a box on graph paper. They then test this box by trying to visualize how many of each package will fit.

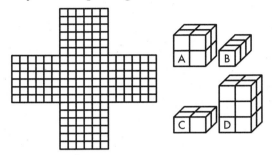

Using Numbers Some students generate possible dimensions for boxes by thinking about numbers. They are exhibiting an emerging knowledge of the concept of numerical factors.

Marcus proposes that a box 12 by 2 by 2 might work.

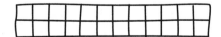

Marcus *[moving a single package A along a 12 × 2 rectangle he has drawn on graph paper]*: 2, 4, 6, 8, 10, 12. That fits. Now, for package B—3 fits into 12. That's 3, 6, 9, 12. *[He then checks his answer by moving package B's along a 12 × 1 strip on his rectangle.]* That's one [12 × 1] row, then I could put another row here *[beside it],* and here and here *[above the previous 2].* And package D works because I just showed that 3 fits into 12. And C works because there are 3 C's in a D.

Notice that as this episode progresses, Marcus relies more and more on manipulation of numbers rather than packages: "D works because I just showed that 3 fits into 12." The connections he is establishing between numerical and spatial ideas lay an excellent foundation for the concepts of factors and divisibility.

More Packing Problems

What Happens

Students predict how many two-cube packages fit into boxes that are marked off by rectangles, not squares. This activity uncovers a misconception many students have about multiplication and arrays. Student work focuses on:

■ figuring out exactly how to use multiplication to predict the number of two-cube packages that fit in boxes

Materials

■ Student Sheet 7 (1 per pair)

■ Interlocking cubes (70 per pair)

■ Scissors and tape for each pair

■ Student Sheet 8 (1 per student, optional)

Activity

Distribute Student Sheet 7, More Packing Problems, to each student. Students predict how many two-cube packages will fit in box 1, then describe in writing how they found their answer. Working in pairs, the students make box 1 from the pattern, then check their answers with packages of cubes. They each record the actual answer on their student sheets.

Assessment

More Packing Problems

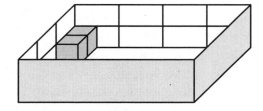

Follow the same procedure for box 2, with students predicting individually, but checking their answers as a pair.

There is a common error that students make in doing these and similar problems. Because of the way the side walls are marked, to find the number of packages in a horizontal layer, they will, for box 1, simply multiply 3 by 4. Of course, if they make the box and fill it with cube packages, they will see that 24 (not 12) packages fit in the bottom layer. See the **Teacher Note,** A Common Misconception (p. 51), for a full discussion of students' strategies and this error.

As you walk around the room during small-group work, you will notice students making this error. If they do not see the source of the problem, it is important that you bring it to their attention.

Why did you multiply 3 times 4? (A student might answer, "Because there are 3 rows of 4.") **So why didn't this work—why didn't it give you the correct answer?**

After students have completed both problems, have them discuss their solutions and strategies in a whole-class discussion. Be sure to have several students explain why the above-mentioned common errors occurred.

How many of you predicted 24 packages for box 1? How did you make your prediction? What was the actual answer? Why didn't your prediction strategy work—why didn't it give you the correct answer?

Collect the student sheets. Reading these will give you a good indication of individual students' progress on this type of problem. Expect that many students will make the error discussed above when doing the first problem, but far fewer should make it on the second.

Session 5 Follow-Up

 **Extensions**

- **Economical Boxes**   Student Sheet 8, Saving Cardboard, takes the packaging idea in a different direction. The task here is to design boxes that will hold 8 cubes, then determine which of these requires the least amount of cardboard. If they have time, they might also design the most economical box that will hold 24 cubes. These problems prepare students to distinguish between volume and surface area.

- **Student-Generated Problems**   In a few classrooms, students have spontaneously designed their own problems to challenge their classmates. Encourage all students who take this initiative. However, you may need to provide some guidance to keep the problems manageable. For example, if students want their classmates to determine the number of cubes that will fit in a familiar space in the classroom, suggest a relatively small space that is rectangular or square, and continue to make interlocking cubes available.

A Common Misconception

When doing the first problem on Student Sheet 7, many students look at the rectangles on the sides and incorrectly think that there are 3 times 4 packages on the bottom of the box. They don't correctly visualize how the packages will actually fit in the bottom layer of the box.

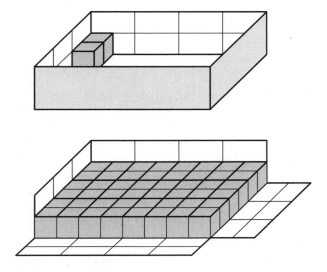

This error is a symptom of students' incomplete understanding of multiplication. These students think that multiplying the length times the width gives the number of *objects* in the array. But they don't organize the objects into rows and columns so that the width gives the number of cubes in a row, and the length the number of rows (or vice versa). Instead, they unthinkingly mulitply the two most obvious numbers. It is essential to encourage students to actually visualize the way packages fit in the boxes, as the following examples illustrate.

Trevor: For the bottom, there's 4 times 3, equals 12. Divide that by 2 because there are 2 cubes in a package; 6. And there's 2 layers. So it's 12.

Natalie: At first we counted each cube and came up with 48 [for the bottom layer]. But we remembered that each package has 2 cubes. So we divided 48 by 2 to equal 24 packages on the bottom; 24 more on top makes 48 in all.

Trevor and Natalie both recognize that they have to adjust their counting strategy because there are 2 cubes per package. Natalie determines how many unit cubes fit in the box, then correctly divides by 2 because there are 2 cubes per package. Trevor's division by 2, however, is inconsistent with the actual way the packages fit within the box.

Jasmine: There's 3 in a row along the side. There's 8 along the back—because 2 fit in each space. So there's 24 on the bottom, plus 24 more on top; 48.

Mei-Ling: There's 6 on the left side [top and bottom], and 8 rows, so 48.

Jasmine and Mei-Ling both correctly imagine how the packages fit within the box, permitting them to find the correct number of packages.

Measuring the Space in Our Classroom

What Happens

Session 1: Measuring the Space in a Box
Student pairs determine the number of cubic centimeters that fill an unmarked, closed box that measures 5 by 6 by 8 cm. During a class discussion, the term *volume* is introduced. Students then decide on a unit of measure for finding out how much space is in their classroom and construct a model of this unit.

Session 2: The Space Inside Our Classroom
Student pairs determine the number of cubic meters that fit inside their classroom. They describe and justify their methods and discuss discrepancies in answers.

Session 3 (Excursion): Measuring the Space in Other Rooms Students determine the volume of another room in the school to compare its size to that of their classroom. They write a description of the method they used and tell why they think it is valid.

Mathematical Emphasis

■ Understanding the concept of volume and units of volume

■ Seeing cubic centimeters as a unit for measuring volume

■ Deciding on, constructing, and visualizing appropriate units of volume for measuring a large-scale space, such as a classroom

■ Understanding characteristics of units of volume, such as shape and size

■ Developing meaningful methods for determining the number of volume units that fit in a solid shape—that is, methods that can be visualized, explained, and justified

■ Describing and justifying methods of determining volume

■ Comparing the volume of one room to that of another room

What to Plan Ahead of Time

Materials

- Scissors and tape for each pair (Session 1)
- Centimeter rulers: 1 per pair (Session 1)
- Centimeter cubes: 100–150 (not all students will use them, but they will be helpful for some) (Session 1)
- Calculators: available (Session 1)
- Metersticks: 1 per pair. If metersticks are in short supply, you will need from 3 to 12 sticks in one-meter lengths for building a cubic meter model in Session 1 (see p. 58). Students could use more such sticks or string cut into one-meter lengths for measuring tools in Sessions 2 and 3.
- Yardsticks: 12 to make a cubic yard model, if available (Session 1)
- Foot rulers: 12 for each cubic foot model, or sheets of cardboard larger than 12×12 (Session 1)
- Masking tape: 1 roll for each group of 4–6 students (Session 1)
- Interlocking cubes: available (Sessions 2–3)

Other Preparation

- From the pattern on Student Sheet 9, build a demonstration closed box to show when you introduce the problem. Also try the problem yourself before Session 1.
- During Session 1, students will be building models of standard volume units (see p. 57 for details). Determine in advance how many cubic meters, yards, and feet you have the materials to build so that you can group students accordingly for this activity.
- Before Session 2, out of sight of your students, measure the dimensions and find the volume of your classroom in cubic meters, so you can judge about how accurate your students' answers are.
- Duplicate student sheets and teaching resources (located at the end of this unit) as follows:

For all sessions

One-centimeter graph paper (p. 156): a small supply for use as needed

For Session 1

Student Sheet 9, Pattern for a Closed Box (p. 132), 1 per pair

Measuring the Space in a Box

Materials

- Demonstration model of closed box
- Student Sheet 9 (1 per pair)
- Scissors and tape for each pair
- Centimeter rulers, centimeter cubes, and calculators (available for use as needed)
- Metersticks, yardsticks, twelve-inch rulers (or cardboard larger than 12 × 12), and graph paper, for model units
- Masking tape (1 roll per 4–6 students)

What Happens

Student pairs determine the number of cubic centimeters that fill an unmarked, closed box that measures 5 by 6 by 8 cm. During a class discussion, the term *volume* is introduced. Students then decide on a unit of measure for finding out how much space is in their classroom and construct a model of this unit. Student work focuses on:

- determining the volume of a small box in cubic centimeters
- understanding the notion of volume and units of volume
- deciding on, constructing, and visualizing appropriate units of volume for measuring a large-scale space, such as a room
- understanding characteristics of units of volume, such as shape and size

Activity

Finding Cubic Centimeters Without Cubes

Show a centimeter cube as you introduce the first activity.

This cube measures 1 centimeter along each edge. It is called a *cubic centimeter*. Today you're going to figure out how many cubic centimeters it will take to completely fill this box. [*Show the closed box you have prepared.*] **Each pair will get a pattern to make a box like this. You will have scissors and tape, and you can also use centimeter rulers and calculators.**

Because we don't have a large supply of centimeter cubes, you need to come up with some measuring strategies that don't depend on having lots of cubes. For that reason, see if you can figure out how many centimeter cubes fit in the box *without* using cubes. When you have an answer, write it down. Then we'll talk together about what you found.

Distribute Student Sheet 9, Pattern for a Closed Box. As students cut out and tape the box together, *do not* tell them the dimensions of the box (5 by 6 by 8 cm). Make available centimeter rulers, centimeter cubes, and calculators for use as needed.

As students work, move from pair to pair and observe their strategies. If students have measured the dimensions, ask them why they did so. If students don't specifically say they measured the dimensions to find out how many cubes fit along each edge, ask them:

How many cubes will fit along one edge of this box?

Their answer will tell you whether they understand the significance of their ruler measurements. If they don't know, give them a cubic centimeter and repeat the question.

For students who can't get started on this problem, try giving the following hints. Starting with the first hint, try giving one at a time until students are able to work on their own.

- **Could using a ruler help you?**
- **What if you measure the edges of the box—will that help?**
- **What if we mark centimeters off on the edges?** *[You might have to illustrate this for students.]*
- *[Show a cube.]* **Here is a cubic centimeter. Can it help in any way?**
- *[Point to one edge of the box.]* **How many cubes fit along this edge?**
- *[Offer graph paper with a one-centimeter grid.]* **If you make the box from this paper, does that help?**

Students may check their answers by using a few centimeter cubes if they wish.

The goal of these hints is to help students understand that measuring the dimensions of a box in *length* units indicates the number of *cubic* units that fit along the edges of the box. Some students do not understand this relationship.

Suppose, for example, that students measure the dimensions of the paper box as 5, 6, and 8 centimeters. Some will not understand that this means 5 cubic centimeters fit along the bottom front edge, 6 cubic centimeters fit along the right vertical edge, and 8 cubic centimeters fit along the depth. Other students either will not think of using centimeter rulers to measure the box, or will feel uncomfortable doing so because they don't understand this relationship.

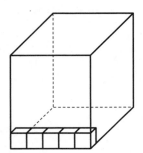 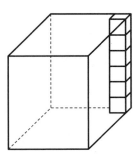 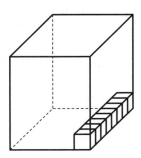

Sharing Strategies with the Class   After most students have found a solution, students share their strategies for determining the number of cubic centimeters that fit in the closed box. Ask several pairs of students who had *different* strategies to explain the following:

How did you find the number of cubic centimeters that fit in the box? Why did you use that method? How do you know your method is correct?

If a number of students were unsuccessful in solving the problem, match up pairs who had difficulties with those who were successful. This sharing time is an opportunity for students to help each other so that everyone will have seen a useful strategy by the end of class.

Activity

Units of Volume

If students have already worked in the grade 5 *Investigations* unit, *Measurement Benchmarks,* they are familiar with liquid volume as the amount a container holds, measured by such units as liters and milliliters (metric system) or quarts, cups, and fluid ounces (U.S. standard measure). Here, they expand their understanding of the term *volume* as they measure space with cubic units. For this discussion, have at hand the closed box and the centimeter cubes from the preceding activity.

This is the closed box you just worked with. We call the amount of space inside this box its *volume*. The volume of a three-dimensional object is the amount of space enclosed by its outer boundary. That means even a solid object has volume. For example, this chalkboard eraser is solid, but it still has a volume.

When we measure length, we use units like meters, and what else? (centimeters, millimeters, kilometers; inches, feet, yards, miles)

And when we measure weight, we use units like what? (kilograms, grams; pounds, ounces)

We also use special units to measure volume. For example, we could use cubic centimeters *[hold up one cube]* to measure the volume of this small box. We could say that the volume of this box is 240 cubic centimeters, because it takes 240 of these cubes to fill it.

What are some other units of volume?

List students' ideas on the board, or help them out as necessary by suggesting units such as cubic meters, cubic inches, cubic feet, and cubic yards. As needed, illustrate the difference between units of volume and length by comparing an inch on a ruler to a cubic-inch block, or a centimeter length on a ruler to a centimeter cube.

What do all these volume units have in common?

Help students see that they are *cubes,* and the lengths of their edges are commonly used units of length.

Some students may wonder why we use cubes for measuring volume. They may think, for example, that the concrete blocks or bricks used to construct their classroom walls could be used for volume units. They are correct, but there are several disadvantages to using these units—which you can discuss if students suggest that idea.

First, remind students of the difficulties they encountered with the packing problem in Investigation 2 that asked about the number of *two-cube* packages that fit in a box. It wasn't as easy to count these packages as it was to count single cubes. That's a major reason to use cubes to measure space: they are easier to count. Because they are the same length along each edge, we don't have to worry about how to orient them.

The second disadvantage of using rectangular blocks or bricks is that they are not a standard size. We may not even know the exact shape of the blocks, because we can't see their depth. To measure the space inside a room *without* these blocks in their walls would be very difficult; cubic feet or cubic meters give us a standard unit to use in *any* room.

Building Models of Volume Units

Explain to students that in the next session, they are going to be measuring the space inside their classroom. That is, they will be finding its volume.

What standard volume units do you think would be best to use for measuring the volume of our classroom?

In this activity, the class will build models of at least five standard volume units, including both U.S. standard and metric units: cubic centimeter, cubic inch, cubic foot, cubic yard, and cubic meter. Seeing models of these different units of volume will help students think about which would be an appropriate unit for measuring the volume of the classroom.

Group students according to how many materials you have available for the different units. For building cubic yards and cubic meters out of sticks (as described below), plan on groups of four to six for each unit, as it takes this many students to hold the sticks in place while they are being taped. Groups of three to four can build cubic feet using rulers (following the same procedure as for metersticks) or sheets of cardboard cut and taped together. All remaining students can work in pairs to make cubic inches and cubic centimeters with paper, scissors, and tape.

Building a Model Cubic Meter   Students can build a cubic meter using 12 metersticks, joined at the corners with plenty of masking tape. If you have more wall space than metersticks, students can create a model of a cubic meter by marking out a square meter on the floor with masking tape, and another square meter in masking tape on the adjacent wall. Adding five metersticks for the top and outside edges will complete the model. If you have an available corner, the same general plan will work with only three metersticks.

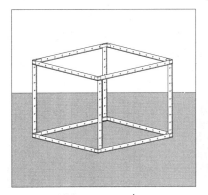 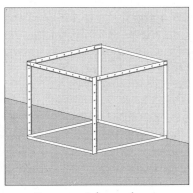 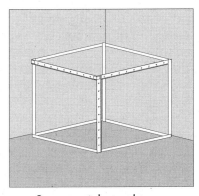

| 12 metersticks freestanding | 5 metersticks and tape against wall | 3 metersticks and tape in corner |

Follow the same approaches for building a cubic yard and a cubic foot.

Discussion: Which Unit Should We Use?   After students have finished building their models, display the units where everyone can see and compare them. The **Dialogue Box,** Talking About Units of Volume (p. 62), demonstrates some of the terminology problems students have in talking about the measure of volume during this discussion.

Remind the class that in the next session, they are going to be measuring the space inside their entire classroom.

Which of these units do you think would be best to use for measuring our classroom?

See the **Dialogue Box,** What Kind of Units? (p. 63) for some examples of students' thinking about this issue.

Suppose we measured first with a small unit—say, cubic centimeters—and then with a bigger unit, like cubic meters. How would the number we get for the amount of space in the room be different? Are there more cubic centimeters or cubic meters in the classroom?

Students should recognize that measuring with a larger unit is more manageable; the numbers stay smaller, and the actual measuring is easier.

The students may want to talk about their ideas for how to find the number of units in the classroom. While they may need to do some of this in order to explain why they think their unit will work or is best, try to avert such discussion because pairs will be devising their own plans for measuring the classroom in the next activity.

Cubic feet, cubic yards, and cubic meters are all a good size for measuring the classroom. Everyone needs to use the same unit so that we can compare our volume measurements when we are done. For this activity, we will use cubic meters.

Save at least one cubic meter that students constructed so that you can refer to it during the next two sessions as students discuss how many cubic meters are in their classroom.

A Plan to Measure Classroom Space

In the next session, students will extend the schemes that they developed with small cubes and boxes to determine the amount of space in—or volume of—their classroom. To solve this problem meaningfully, students will need to *mentally* construct cubic meters and visualize how these units fill the classroom.

If your classroom is not a rectangular prism, make the problem manageable by specifying some part of it that is. Also, to simplify the calculations, students should round to the nearest whole unit.

Imagine our classroom completely empty—no desks or chairs or shelves, just the floor and walls and ceiling. How much space for air is there in the empty classroom?

You're going to make a plan for measuring the amount of space in the large "box" that is our classroom. Your plan should work for any room. That way, you can easily find and compare the amount of space in two rooms.

Because we are going to use cubic meters as our unit of volume, the problem is to find how many cubic meters fit in our classroom. You can use metersticks, string, calculators, and any other tools we have.

Initially, some students might not focus on volume, taking the area of the floor instead of volume to quantify the space in the room. Although this is not the point of the activity, it is consistent with everyday conceptions of the size of a room. For example, for real estate and architecture purposes, the size of a house is usually given in terms of its floor space—that is, the total area of the floor in square feet. Often, when we think about the amount of space in a room, we're really wondering how many people or objects will fit on the floor. If you find students thinking in terms of floor area, emphasize finding the amount of *space for air* in the room.

Are we going to put air only on the floor? Or will air fill every part of the room, even up to the ceiling?

If very many students are thinking about area instead of "space for air," discuss this as a whole class. If necessary, compare the model cubic meters with a square meter (laid out, for example, in masking tape) and with a meterstick. Ask students to think about the different things these three different units (meter, square meter, cubic meter) can measure.

Working in pairs, students write their plans for measuring the volume of the classroom. In the next session, they will implement their plan and compare their results with other students.

❖ **Tip for the Linguistically Diverse Classroom** Pair students so that those who are proficient in English can do the writing, while those who have limited English proficiency can add drawings or diagrams to help clarify the written plan.

> We are going to mesure the three dimensions and multiply length and width and then multiply the anwser times the height. We are going to mesure with meter sticks.

> measure a cube. Then take cubes and measure from the floor to the ceiling. Keep the number and then measure from East to West. Then measure from North to South. Multiply them all and then you will find out how many cubes would fit in the classroom.

Session 1 Follow-Up

Homework

Start gathering materials for Investigation 4. Ask students to bring from home three or four empty and clean containers, such as small bottles from mustard, jam, or spices, small boxes from toothpaste or soap, and tin cans. As a general guideline for size, tell students that three of the containers should fit comfortably in a paper lunch bag. Students will be pouring rice or sand into the containers to compare their volumes. Therefore, bottles should not have small openings, and boxes should have sealed edges. The family recycling bin may be a good source for these objects.

Talking About Units of Volume

After the activity Building Models of Volume Units (p. 57), these students are considering the different volume units they have built and gathered. The students are having difficulty with the names of their units. Through discussion, the teacher helps them understand that each of the volume units are cubes. What distinguishes them is their size, which has to be described whenever one of the cubes is mentioned.

What is a cube?

Natalie: A square!

A square?

Natalie: It's like a dice. It's got six sides.

Anything special about those six sides?

Natalie: It's like if you put square blocks on them.

What makes this a cube? Or what doesn't make it a cube?

Jeff: On that *[pointing to the cubic foot],* the sides are all equal to each other, and on that *[indicating a cement block in the classroom wall]* they aren't.

[The teacher holds up, for comparison, a square made from four 12-inch rulers and a cubic foot.] **How would you describe these?**

Toshi: One's a square, and one's a bigger square [the cube].

Toshi, how is a cube different from a square?

Toshi: This one [the cubic foot] you can see everything inside of it. It is more like a cube. And that [the square foot] is more like an outline of it [cubic foot].

So, is this *like* a cube, or *is* it a cube?

Toshi: It *is* a cube.

If you wanted to talk about these with somebody else, what would you call them?

Corey: That's a square cubic foot and that's a square foot.

Manuel: If I wanted the three-dimensional square, I'd say, can I please have, um, a cubic foot. If I wanted this *[holding up the square],* I'd ask for the two-dimensional square.

Jeff: Well, this should be called a cube, and this should be called just a square. Because you know how you have those little cubes of ice, you don't call those squares, you call them cubes.

[The teacher now stands next to a cubic meter and cubic yard that students have made. She places the cubic foot, a cubic inch, and cubic centimeter where all students can see them.]

Would everybody agree that these are all cubes?

Students *[variously]:* Yes. That's right.

How are these cubes different? How can you tell somebody which cube you are using?

Julie: Well, that one is 1 foot. That one is 1 inch.

So how should I name these different cubes when we are talking about them?

Christine: We should tell how big they are. Like a foot cube, or a meter cube.

So to help us understand what cubic units we're talking about, we agree to tell how big they are, the way Christine did. Mathematicians usually call a foot cube *[pointing to the model]* a cubic foot, and a meter cube a cubic meter. But we can use either term in our class—foot cube or cubic foot—and understand what we are talking about.

What Kind of Units?

These students are discussing which would be the best unit for measuring the volume of a large area—their classroom (A Plan to Measure Classroom Space, p. 59).

What would be good units to use for measuring the amount of space in the classroom, and why?

Matt: Centimeters.

Becky: Feet—they're bigger. You wouldn't have to count up all the centimeters.

Amir: Any metric, really. Like decimeters, because it makes the numbers smaller.

Sofia: Yards. It's just easier—you'll get smaller numbers.

Yu-Wei: Meters—they're bigger than yards so you get even smaller numbers.

Robby: I think it's harder to use big units like meters because if it's half, you don't know what to do with it. Smaller units are more accurate.

Heather: We were going to figure out how many blocks [from the walls] are in the classroom, then figure out how many cubic inches fit in each block.

What could be some problems in using blocks?

Robby: The block on one wall takes up one block, but on the other wall it takes up two. If you count blocks on this wall, there would be two times as many. The blocks are rectangles.

Alani: It would be a lot easier to use square cubes, not rectangles.

These students have a good understanding of what factors might influence which units they should use. They understand that smaller units of measure give larger numbers for measurements. They are also able to suggest the possible pitfalls of the method Heather proposed—using blocks to find the number of cubic inches in the room.

The Space Inside Our Classroom

Materials

- Metersticks or string cut in one-meter lengths (1 per pair)
- Graph paper (available)
- Interlocking cubes (available)

What Happens

Student pairs determine the number of cubic meters that fit inside their classroom. They describe and justify their methods and discuss discrepancies in answers. Student work focuses on:

- understanding the notion of volume and units of volume
- developing meaningful methods for determining the number of cubic meters that fit in a room, that is, methods that students can visualize, explain, and justify

Activity

How Many Cubic Meters in Our Classroom?

We are trying to find out how much space for air there is in our classroom. Yesterday, we decided to do this using cubic meters as our volume units, and you worked on your plans.

Today you will carry out your plans to find the number of cubic meters that fit in the classroom. You and your partner should record your answer to this problem and explain how you found your answer. You must be able to convince your classmates that your answer and the method you used are correct.

Students continue to work in pairs as they determine the amount of space in the classroom in cubic meters. Remind them to round their measurements to the nearest whole unit. Thus, if they first measure the length of the classroom, that measurement should be rounded to the nearest meter. As students come up with answers, they might check with other pairs. If they didn't get the same answers, they should try to determine why.

Pairs will use a variety of strategies, some of which they might not fully understand. For example, some students measure the dimensions of the classroom in meters, but they do not understand what these measurements indicate about how cubic meters fit in the classroom. See the **Teacher Note,** Strategies for Measuring Space in the Classroom (p. 70), for examples of other general strategies students may use.

If students are having difficulty, try the following:

- Ask questions that help them see the relationship between length measurements and volume.

  How many cubic meters do you think will fit on the floor, along this wall? How could we find out? How many rows of these cubic meters will cover the whole floor? How many layers could we stack to reach the ceiling?

- Suggest that students model the situation with interlocking cubes or paper boxes. They could make a scale-model classroom, either as a paper box from graph paper or as a solid package of cubes, each cube representing a cubic meter. Thus, if the room measures 10 by 8 meters on the floor and is 3 meters high, students would make a box or package that measures 10 by 8 by 3 cubes. This model should help students better visualize how cubic meters fit into the classroom.

- If students are still having difficulty, suggest they work first on a simpler problem. For example, how many cubic feet would fit in a box the size of your desk? This reduces the scale of the problem, making it easier to relate to their earlier work with small cubes and paper boxes.

- Ask questions that encourage students to relate the ways they found volume on a small scale in Investigations 1 and 2 to how they might find the amount of space in large-scale environments.

  How did you find the number of cubes in a box? Does that type of thinking help here (for finding the amount of space in a room)?

While these suggestions may be necessary for some, avoid taking a directive approach with the whole class; and give all the students plenty of time to struggle with the problem themselves.

Justifying Our Methods As students are working, ask questions that help them elaborate and justify their ideas. For example, if they decide to measure the room in meters and multiply these dimensions, ask questions that ensure that they understand what they are doing.

What does your unit for measuring amount of space look like? You measured in meters, but you got an answer in cubic meters. How does that work? When you measured, you found that the length of the room was 8 meters. What did that tell you about cubic meters?

Once students understand the significance of the linear measurements they are taking, they can use this knowledge to apply techniques they developed in Investigations 1 and 2 for finding the number of cubes in three-dimensional arrays. For example, suppose that students have multiplied 8 m by 10 m by 3 m.

Why did you multiply these numbers together? What did 8 times 10 give you? (the number of cubic meters that cover the floor, 80.) **Why did you then multiply by 3? What did that give you?** (There are 3 layers of 80 to fill the whole room.)

Note: While students are measuring the room, furniture and other physical obstacles can pose nontrivial problems for some of them because they don't see alternate ways to find the lengths they seek. Help them find a way to deal with these problems. For example, one pair of students was trying to measure the length of a wall along the floor. Because there was a short bookcase in the way, they were stumped. Until their teacher asked them about it, they didn't see that they could measure along the wall above the bookcase to find the same length.

Discussion: How We Measured

Start the discussion by asking questions that encourage students to review and clarify the problem that was posed. It's important to repeatedly refocus students' attention on *visualizing* what is happening as they measure space.

What was the problem we were trying to solve? (to find the number of cubic meters that fit in the classroom) **What exactly does this number tell us?** (how much space or volume our classroom has) **When you say that there are 240 cubic meters in the classroom, what does that mean?** (that 240 cubes like the one built from metersticks and tape could be put in our classroom)

After reviewing the problem that they have solved, students report their numerical answers. As they report, record the numbers on a line plot.

```
                    X
                    X
          X         X                   X
          X         X                   X
    X     X    X    X    X    X          X
   220   230  240  250  260  270  280  290
```

Ask pairs with discrepant answers to explain and justify their solutions. Answers will differ because of computational errors, incorrect methods, or errors in measuring the dimensions of the classroom. Help students see the different sources of errors as they listen to each other's explanations and justifications of answers and methods.

These two pairs both measured the room's length, width, and height, then multiplied these numbers together. So how could they have different answers? (Computational error or inaccurate room measurements are likely causes.)

Finally, ask students to explain the different methods they used, both in taking measurements and doing the computation. Encourage them to elaborate on what kinds of measurements they used and why.

What parts of the room did you measure? Why? How did you find the number of cubic meters that fit along this edge of the floor? Some people measured one edge of the floor with a meter tape. What did they learn from that?

Be sure that students describe and justify any computations they did. For example, if they say that they multiplied, ask why. Some students might say that they multiplied the three dimensions (length, width, and height) because that's the method they discovered for finding cubes in paper boxes. These students may or may not be able to explain why this procedure works. Other students may say that they determined how many cubes fit along the length and along the width of the room (on the floor). They multiplied these numbers together. Why? To get the number of cubic meters that covered the floor. They then multiplied by the number of cubic meters that fit along the height. Why? Because that's the number of layers. This is excellent reasoning.

Session 2 Follow-Up

Homework

- Students choose a room at home and determine its volume in cubic meters. They will need some measuring instruments, such as meter-long tapes or strings. They are to determine the volume of the room and describe *as well as justify* their method in writing.

  Checking this homework can give you an indication of your students' understanding of the concepts in this investigation (keeping in mind that some students may get help from family members). If they multiplied the three dimensions, how did they justify that this method was valid? Did they think in terms of layers?

- With family help, students find recorded measurements of volume on household items, in manuals, or in other written material. For example, they might find measurements of the amount of trunk space in a car, the amount of water used as reported on the monthly water bill, and the volume of refrigerators and freezers. In some areas, household waste is measured by volume to account for the use of disposal space. Students look for any measurements given in cubic units, recording both the numbers and the kind of cubic units used, to report to the class.

Extensions

■ **How Many Cubic Centimeters in a Cubic Meter?** There are 1 million cubic centimeters in a cubic meter, and this number fascinates students. This is one of the few situations in which students can come to appreciate the magnitude of this often-mentioned number, a million.

This problem is not meant to be a typical "units conversion" problem. By placing plastic cubic centimeters within the cubic meter that they have built, students can visualize a method for solution. Usually there are gasps of surprise when students find that the answer is 1 million. If students need help, you might ask questions like these:

How many cubic centimeters fit along the bottom edge of the cubic meter? (100) How many fit along the other bottom edge? (100) So how many cubic centimeters will it take to fill the bottom layer of the cubic meter? (100 × 100 = 10,000) How many layers of cubic centimeters are there? (100) So how many total cubic centimeters will fit in the cubic meter? (100 × 10,000 = 1,000,000)

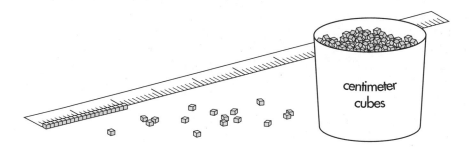

centimeter cubes

■ **Volume with Fractional Measures** Pose this problem:

What is the volume of a room that measures 4 by 6 by 2½ meters?

When they measured their own classroom, students were told to round their measurements to the nearest meter. In this problem, students have to deal with a fractional length. Many will suggest that we can simply multiply the three dimensions to find the answer. Of course, they are correct; but they need to think about why this still works, and to visualize the solution.

Ask students to prove that their method and answer are correct. One approach they might take is to make a model of the room as a paper box (cut from graph paper) and fill it with interlocking cubes (each cube represents a cubic meter). After completing the box, students will see that 24 cubes fit on the bottom layer, and another 24 cubes fit on the second layer. But whole cubes will not fit on the next layer. In fact, if they place another layer of 24 cubes in the box, only half of each cube will actually remain in the box. So, if they could cut each of these cubes in half, they would completely fill the box. They would need ½ of 24 cubes to fill the top layer. So the volume of the room is 60 cubic meters.

■ **The Volume of Large Classroom Objects** Students might determine the cubic meters of volume for objects such as a desk, a filing cabinet, or a bookcase. Some of these objects present a real challenge. For example, a cubic meter will not fit in most bookcases. So, how can we figure out the number of cubic meters of space in the bookcase?

One method is to use fractions or decimals. For example, if the bookcase measures $\frac{1}{3}$ m in depth, 1 m in width, and 2 m in height, students might compute its volume as $\frac{2}{3}$ of a cubic meter. Ask them if this makes sense. Have them try to visualize cutting the bookcase into pieces and fitting these pieces inside the cubic meter that has been built. How much of the cubic meter of space is filled?

If students use decimals, help them see that centimeters are hundredths of a meter. They might measure the bookcase described above as 0.33 m by 1 m by 2 m, getting a volume of 0.66 cubic meter.

Getting into fractional parts of cubic units is a difficult problem that many students won't be able to handle, but some will find it an interesting challenge.

Teacher Note

Strategies for Measuring Space in the Classroom

Students can meaningfully determine how many cubic meters fit in their classroom by visualizing the repeated placement of cubic meters throughout the room. For example, they might repeatedly place a single meterstick end to end along the wall, imagining the cubic meters projecting out from the wall. Be sure to ask students what they are doing; try to see if they are imagining cubes or simply measuring the wall's length. Other students might actually want to repeatedly place a cubic meter model on the floor along one of the walls, making the row of cubes more concrete.

The examples below illustrate the complexities that students often introduce into this task.

Converting Cubic Centimeters to Cubic Meters   Tai and Danny use the concrete blocks in the walls to correctly find the number of centimeters in the length, width, and height of the classroom. They reason:

> 1 block = 21 cm high, the height of the room is 15 blocks, so the height of the room is $21 \times 15 = 315$ cm.

> 1 block = 40 cm long, one wall is 28 blocks long, so the length of this wall is $40 \times 28 = 1120$ cm.

Similarly, they find the width of the room to be 810 cm. They then multiply the three dimensions together on a calculator, getting 285,768,000.

Danny: To get cubic meters, just divide our last answer by 100.

Tai: Why divide by 100?

Danny: Because there's 100 centimeters in a meter.

After overhearing the boys' conversation, the teacher takes several cubic centimeters and drops them into the meterstick-model of cubic meter as the boys watch.

Do you think 100 of these [cubic centimeters] fill one of these [the cubic meter]?

Tai and Danny: No!

How about converting your original measurements to meters?

Tai: I think that would be better.

The boys then correctly change each of their original measurements into meters and multiply them to find the number of cubic meters in the classroom. If the teacher had suspected that they would have had difficulty converting meters to centimeters, she could have had them remeasure the room in meters, rather than centimeters.

As almost always happens with students of this age, when Tai and Danny tried to convert one volume unit to another, they thought only about the relationships between the corresponding *length* units. They did not properly visualize how many *cubic* centimeters fit in a *cubic* meter. The teacher wisely redirected the students from trying to convert cubic centimeters to cubic meters to directly determining the volume of the room in cubic meters.

Converting Blocks to Meters   Jeff, Trevor, and Maricel find the dimensions of the room in meters by noticing that there are 5 horizontal blocks ("bricks") for each 2 meters. They count the number of blocks in the length and width, then convert these block measurements to meters.

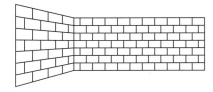

Jeff: [For the length] $30 \div 5 = 6$, $\times 2 = 12$ meters. [For the width] 20 bricks $\div 5 = 4$, $\times 2 = 8$ meters. So $8 \times 12 = 96$ square meters on the floor.

The students find that 5 blocks high make 1 meter, and determine from this that the height of the room is 3 meters. They multiply the dimensions together and get 288.

Continued on next page

What are the units?

Jeff: Meters.

Trevor: Square meters.

Maricel: Cubic meters.

Jeff and Trevor: Right, cubic meters.

Do you think that 288 of those cubic meters [pointing to the cubic meter model] **will fit in this room?**

Jeff: That's an awful lot.

Maricel: I don't know.

Why don't you make a model of the classroom to see if your answer makes sense.

The students decide that one interlocking cube will stand for one cubic meter, then they build the four walls of the classroom with cubes. This model convinces them that they are correct.

Jeff: Look how many cubes we have already and we don't even have the middle filled.

I know another group got 240 cubic meters. What do you think?

The three students decide that maybe somebody measured wrong. They ask the teacher what the other group's measurements were.

They measured the room as 3 by 8 by 10 meters.

Maricel: It's either that we counted wrong, or that they counted wrong.

The students recount the blocks several times getting a variety of answers 28, 29, 30. Finally, they recount the blocks together very carefully, getting 30.

Maricel: The other group counted wrong. We recounted a zillion times. We think the length is 12 meters.

Jeff, Trevor, and Maricel did some excellent thinking. But notice that their solution path was indirect; they had many questions to resolve:

- They used blocks in the walls to find the number of meters, rather than measuring in meters directly.

- They had to think about what the unit of measure was called.

- They needed to make a model to convince themselves that the magnitude of their answer was reasonable.

- They had to remeasure the length several times to convince themselves that their answer was correct, and that the other students' answer must be wrong.

This struggle to make personal sense out of their procedures was essential for these students. The teacher asked just the right questions to encourage them to reflect more on what they had done and clarify their thinking.

Measuring the Space in Other Rooms

Materials

- Metersticks or string cut in one-meter lengths (1 per pair)
- Graph paper (available)
- Interlocking cubes (available)

What Happens

Students determine the volume of another room in the school to compare its size to that of their classroom. They write a description of the method they used and tell why they think it is valid. Student work focuses on:

- understanding the concept of volume and units of volume
- using meaningful methods for determining the number of volume units that fit in a solid shape
- comparing the volume of one room to that of another room
- describing and justifying their methods of determining volume

Activity

Measuring Another Room

As students determine the volume of another room in the school, they reflect on and refine the procedures they invented to find the volume of a large space. Arrange with four or five other teachers to allow your students into their rooms for 10 or 15 minutes to make the needed measurements. Plan to send at least two pairs of students to each room so they can compare answers.

I'm thinking about teaching in a different classroom next year. My choices are *[specify the rooms where students can go to measure]*. **I want to know which of these rooms have the same amount of** *space* **as our classroom. I want to be sure all my things still fit inside. I would like you to help me.**

Choose one of these rooms. Plan how you will find its volume. You will have only about 10 minutes to make any measurements you need. You must then use these measurements to figure out the volume.

As before, students use cubic meters as the unit of volume.

Writing a Report on the Volume

After students have found the volume of their chosen room, they write *individual* reports on their findings. Each student should describe which room was investigated, its volume, what measuring methods were used (including all their measurements and computations), and why the student thinks these methods are correct. Students can talk to their partners, but each must write a separate report. Encourage students to use sketches to explain what measurements they took.

❖ **Tip for the Linguistically Diverse Classroom** Offer students who are not writing comfortably in English the alternatives of doing a pictorial report, which sequences the steps of how they found the volume of their chosen room, or an oral report, with their written measurements and computations for backup.

Sharing Our Findings

When students have completed their reports, hold a class discussion of their findings. Make a chart at the front of the room showing the name of each room (including your classroom). Record students' results for dimensions and volume of each room. Students tell about the methods they used for finding volume. Ask several to explain why they think their methods work, and push them to explore any discrepancies in answers.

What method did you use to find the volume of Room 10? Why did your method work? Did anybody find a room that has the same volume as our classroom? Did this room have the same dimensions as our classroom (length, width, and height)? What rooms had greater volume? What rooms had less volume?

Prisms and Pyramids, Cylinders and Cones

What Happens

Session 1: Comparing Volumes After measuring how much rice or sand small household containers will hold, students order them from least to greatest volume. They begin making geometric solids for use in Sessions 2–3.

Sessions 2 and 3: Comparing Volumes of Related Shapes Students work with 11 geometric solids they make from paper. Each solid is paired with another that has equal base and/or height measurements—pyramids are paired with related prisms, and cones with related cylinders. Using rice or sand, students compare the volumes of the solids in each pair. Finally, as an assessment, students construct a rectangular prism with the same base and three times the volume of a given rectangular pyramid.

Sessions 4 and 5: Using Standard Volume Units With the help of a special measuring tool—a see-through graduated prism—students figure out a method for determining the volume, in cubic centimeters, of each of their 11 solids.

Session 6 (Excursion): How Do the Heights Compare? Students determine what height to make a cylinder so that it has the same base and volume as a given cone. Then they figure out the dimensions of a square prism that will have the same base and volume as a given square pyramid.

Sessions 7, 8, and 9: Building Models As a final project, students design and create a model made from geometric solids. The model must include prisms, pyramids, cylinders, and cones, all of which students can create from patterns. Students describe each solid they use in terms of its dimensions and volume. They estimate and then determine the actual total volume of their model in cubic centimeters.

Mathematical Emphasis

- Comparing volumes of containers of different shapes
- Determining methods for using standard units of volume to measure nonrectangular solids
- Exploring volume relationships between solids with the same bases and height
- Estimating volumes of different solids
- Understanding the structure of solids through building them

What to Plan Ahead of Time

Materials

- Empty containers brought from home: 3–4 per student (Session 1)
- 20–25 paper bags, large enough to hold three small containers (Session 1)
- Rice or sand: ½ pound per pair (All sessions)
- Trays: 1 per pair (All sessions)
- Scissors and tape for each pair (All sessions)
- Centimeter rulers: 1 per pair (Sessions 2–9)
- Centimeter cubes: available (Sessions 4–9)
- See-through graduated prism: 8 per class (Sessions 4–9)
- Liter measuring pitchers: 4 per class (Sessions 4–9)
- Calculators: 1 per pair (Sessions 4–9)
- Overhead projector (Sessions 7–9)

Other Preparation

- Sort the containers students brought from home, putting three with different shapes but of similar volume in each bag. It should be difficult to tell the relative size of the three containers visually. Number the bags. (Session 1)
- Duplicate student sheets and teaching resources in the following quantities.

For Session 1

Student Sheet 10, Pairs of Solids (p. 133) : 1 per pair

Solid Patterns A–K (pp. 139–143), on card stock: 1 per pair. Arrange for pairs to make

these outside of math class before Sessions 2 and 3. Also make a teacher demonstration set.

For Sessions 2 and 3

Student Sheet 11, Pyramid and Prism Partners (p. 134): 1 per student

One-centimeter graph paper (p. 156): a small supply for use as needed

Student Sheet 12, Puzzle Cube Pattern (p. 135): 1 per student, homework

Student Sheet 13, Puzzle Pyramid Pattern (p. 136): 3 per student, homework

For Sessions 4 and 5

See-through Graduated Prism Pattern (p. 146): 1 per small group. Copy on transparency film and assemble, if needed.

For Session 6

Cone, Pyramid, and Cylinder Patterns (pp. 144–145): 1 per pair

One-centimeter graph paper (p. 156): 1 per pair, plus extras

For Sessions 7, 8, and 9

Student Sheet 14, Final Project Tasks (p. 137): 1 per group of 2–4 students

Student Sheet 15, Model Planning Sheet (p. 138): 1 per group of 2–4 students

Dimensions of Solids (p. 147): 1 transparency

Pattern Makers (pp. 148–153): 1 per group of 2–4 students, plus extras of each.

One-centimeter graph paper (p. 156): 1 per group of 2–4 students, plus extras

Comparing Volumes

What Happens

After measuring how much rice or sand small household containers will hold, students order them from least to greatest volume. They begin making geometric solids for use in Sessions 2–3. Student work focuses on:

■ comparing the volumes of containers of different shapes

Materials

- ■ Numbered bags with three containers in each
- ■ Rice or sand, trays
- ■ Student Sheet 10 (1 per pair)
- ■ Solid Patterns A–K (1 per pair)
- ■ Scissors and tape for each pair

 Ten-Minute Math: Guess My Number   Once or twice during the next few days, outside of the math hour, play Guess My Unit, a variation of the Guess My Number activity. You will need the Guess My Unit cards (pp. 157–158), one set per small group. If you have used the grade 5 unit *Measurement Benchmarks,* you will already have sets of these cards.

❖ **Tip for the Linguistically Diverse Classroom**   Suggest that students add a small drawing to each card as a visual reminder of the measurement unit.

Players spread all the cards out in front of them, faceup. The leader secretly picks one of the units, writing the choice on a slip of paper to show later. Players then try to guess the unit selected, taking turns asking yes-or-no questions. For example: Is it metric? Is it a measure of liquid quantity? Is it heavier than a pound? Do people use this unit when they cook?

As units are eliminated, players turn those cards facedown. They continue to ask questions until they feel sure of the answer. To discourage random guessing, a player who guesses a unit must explain what clues led to that guess. This discussion should help students begin to think about what makes a good question.

Students can play in groups of three or four. Each group uses one set of unit cards. They rotate the role of leader.

Comparing Volumes of Containers

Distribute the rice or sand for measuring, with trays for catching spills. Place the bags of containers in a central location. Point out that each bag has a number.

Over the past few days, you've been bringing in empty containers—jars and boxes and cans. I have put three in each of these bags. Each pair takes one bag to start with. Your job is to rank the containers in the bag by their volumes—the amount of space inside—from least to greatest. Start by *predicting* the order, then measure to check. Here's what to do:

1. **On a sheet of paper, write the number of the bag you have selected. Write (or draw pictures) to describe the order that you predict, and tell how you made your prediction.**

2. **Use rice [sand] with the containers to determine the actual order.**

3. **Describe in writing the actual order and how you found it.**

4. **After you complete one bag of containers, try some other bags.**

❖ **Tip for the Linguistically Diverse Classroom** Pair students so that one who is proficient in English can do the writing while one who has limited English proficiency can make supporting sketches to show the order and the procedure for finding it.

As you circulate to watch students work, and later during the class discussion, note students' methods of comparing volumes:

■ Do students use compensation to predict? For instance, they might say things like "This shape is bigger at the bottom; but this other shape is taller, so I think they are the same."

■ Do students compare each pair of containers, or can they draw logical conclusions about the relative sizes? That is, if they know that container X has more volume than container Y, and Y has more than Z, can they conclude that X has more than Z *without* directly comparing X and Z?

■ Do they always use direct comparison, or do they use one container as a unit and count the number of times that this unit fills other containers?

The **Teacher Note,** Students' Methods for Comparing Containers (p. 79), gives examples of some different strategies your students may use.

Discussing Our Methods   After all pairs of students have had a chance to order at least two bags of containers, hold a brief class discussion.

What methods did you use to order the containers from least to greatest volume? Did anyone use a different method?

As students talk about their methods, be sure they show the containers they are talking about as they explain their strategies and thinking. Otherwise it will be almost impossible for others to follow their descriptions.

Making Solids from Patterns

Before starting Sessions 2 and 3, each pair needs to make 11 solids from the Solid Patterns A–K. They can start making these at the end of Session 1, and finish during flexible time (or at home, if they have scissors and tape there). Give each pair a copy of Solid Patterns A–K, preferably on card stock. Also distribute Student Sheet 10, Pairs of Solids, which shows what the patterns will look like when cut out, folded, and taped together. If possible, supply each pair with a shoe box or tub to store their work.

Students use scissors and tape to create the prisms, pyramids, cylinders, and cones from the patterns. To get a sharp crease along the edges of the prisms and pyramids, it helps to draw first along the fold lines, pressing firmly with a ballpoint pen or pencil. Patterns should then be folded toward the scored line. Demonstrate how to make one of the cones or pyramids. Emphasize that when folding two sides together to be taped, these sides should meet exactly and not overlap. Apply tape along the entire length of the sides. Otherwise it will be difficult to keep rice or sand inside the solids.

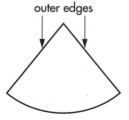  

Note: The patterns for the rectangular prisms (Solid Patterns A and I) are different from the box patterns used in Investigation 1, although they result in similar rectangular boxes. The new pattern is more compact, to better fit on the page. If students seem confused by the new pattern, you might make a box from a pattern in Investigation 1 and demonstrate how it can be cut apart to make a pattern like Solid Patterns A and I.

Students' Methods for Comparing Containers

The structure of the first activity in Investigation 4, Comparing Volumes of Containers (p. 77), is intentionally loose to give students the opportunity to devise their own strategies for predicting and comparing the relative volumes of household containers of different shapes.

Prediction Strategies You'll see a lot of variation in how students make their predictions. One often-used strategy is to visualize or try fitting one container into another, then to compare left-over space.

> This container fits into this one.

> The space around the edges and the heights are the same, but one fits into the other, so that one is bigger.

> If you put this into there, there's room left over.

Another common strategy is to analyze shape characteristics of the containers and focus on particular dimensions to compare.

> These two [cylinders] are the same *[pointing to bases]*, but this one is higher.

> This one is shorter in height, but wider in width.

> This is a lot taller, but this one is a ton wider.

Verifying Predictions To test their predictions, most students directly compare two containers by pouring rice from one container into another. They might pour from a smaller container into a larger and see that the larger is not filled. Or, they might pour from a larger container to a smaller and see an overflow. In either case, they usually draw the correct conclusion.

Another good method is to use one container to judge the other two.

> Two of the small box fill the jar. About 1½ times the large box fills the jar. So the jar is the biggest.

> When we pour the rice from the jar into the toothpaste box, it comes up to here. But when we pour rice from this bottle into the box, it only comes up to here [lower than for the jar]. So the jar has more room.

Students also find less accurate methods.

> We counted the number of handfuls of rice that fit in each container.

These students had a good idea, in that they chose a unit of measure (handfuls) to compare the containers. The problem is that their handfuls were not all equal, so their measurements were quite inaccurate.

Sessions 2 and 3

Comparing Volumes of Related Shapes

Materials

- Student Sheet 10 (from Session 1)
- Prepared solids A–K (1 of each per pair)
- Rice or sand, trays
- Student Sheet 11 (1 per student)
- One-centimeter graph paper (available)
- Scissors and tape for each pair
- Centimeter rulers (1 per pair)
- Student Sheet 12 (1 per student, homework)
- Student Sheet 13 (3 per student, homework)

What Happens

Students work with 11 geometric solids they make from paper. Each solid is paired with another that has equal base and/or height measurements—pyramids are paired with related prisms, and cones with related cylinders. Using rice or sand, students compare the volumes of the solids in each pair. Finally, as an assessment, students construct a rectangular prism with the same base and three times the volume of a given rectangular pyramid. Student work focuses on:

- exploring volume relationships between solids, particularly those with the same base and height

Activity

Identifying the Different Solids

Note: Student pairs should have made all 11 solids before starting Sessions 2 and 3. If they have not, allow time for them to finish before starting this activity.

Small groups show each other their solids and spend a few minutes naming the different types—rectangular and triangular prisms and pyramids, and cylinders and cones. Being able to identify them with correct terminology will be useful for class discussions later in the investigation. Refer to the **Teacher Note,** Geometric Solids and Their Parts (p. 85), for a description of the different types of solids. Help students find their own ways to remember what characterizes prisms, pyramids, cylinders, and cones.

The names of the solids are printed on the patterns to help students identify them. Also, Student Sheet 10 includes names with the pictured solids. Students can use this sheet as a reference throughout the investigation.

Comparing Solids and Their Volumes

Call attention to the way the solids are grouped on Student Sheet 10, Pairs of Solids.

Look carefully at the two solids in each pair. Think about how they are the same and how they are different. Take the solids you have made from patterns and pair them the same way. Use your rice [or sand] to find how the volume of the larger solid compares to the volume of the smaller solid in each pair. I'm going to set up a chart on the board where you can record your findings.

Pair	Solids	Number of smaller solid needed to fill larger solid
1	A, B	
2	C, D	
3	E, F	
4	G, H	
5	I, J	
6	I, K	

Working in pairs, students use whatever methods they choose to compare the volumes of the solids. To contain the rice or sand as much as possible, ask students to do their pouring over the trays.

Note: Because there is likely to be some measurement and construction error in these activities, students won't all get exactly the same answers. Help them recognize when answers are "close enough." Also suggest that they round their answers to the nearest whole or half unit.

If their cutting and taping is fairly accurate, students should discover that the volume of each larger, flat-topped solid is about three times the volume of the smaller, pointed solid, *if it has the same base and height.* (Because of inevitable measurement and construction errors, the 3-to-1 relationship won't be exact.)

Make your chart on the board wide enough for all students to record their answers. Most students will find a 3-to-1 relationship for appropriate pairs of solids. Seeing such numbers recorded in the class chart will encourage those students who find a different relationship to reflect on possible errors.

As you circulate, encourage students to reflect on how the solids in each pair compare. Help them notice that the solids in pairs 1–3 have the same base and height; the solids in pair 4 do not. Similarly, rectangular pyramid K has the same base and height as rectangular prism I, but only the same height—not the same base—as rectangular pyramid J.

Discussion: What We Discovered

After everyone has compared the volumes of the solids in all six pairs, bring the students together to share their findings. The chart you made should remain on the board for this discussion.

What did you discover? How do the volumes of the solids in each pair compare?

How are the shapes of the solids in each pair the same, and how are they different?

The goal is for students to recognize that if a flat-topped solid, such as a prism or cylinder, has the same *height* and *base* as a pointed-top solid, such as a pyramid or cone, then the volume of the flat-topped solid will be three times that of the pointed-top solid. This relationship may be difficult for students to explain fully and precisely. The **Dialogue Box,** Exploring the 3-to-1 Relationship (p. 87), illustrates how some students have described this relationship between the solids.

In order to communicate their findings clearly, students will need to agree on the meanings for the terms *base* and *height* and about methods for measuring the height of cones and pyramids. These ideas are discussed in the **Teacher Note,** Geometric Solids and Their Parts (p. 85). The **Dialogue Box,** Exploring the 3-to-1 Relationship (p. 87), shows how one teacher dealt with these concepts as they came up in the discussion.

As you wrap up this discussion, ask student pairs to save all their solids for use again in Sessions 4 and 5.

Teacher Checkpoint

Pyramid and Prism Partners

Give each student a copy of Student Sheet 11, Pyramid and Prism Partners. Have extra sheets of one-centimeter graph paper available for students who might need them.

In the six solid pairs that you worked with earlier, some of the solids had three times the volume of their partner solid. This sheet gives you a pattern for another rectangular pyramid. First you need to cut out, fold, and tape the pattern to make the pyramid. Then your job is to make a "partner" for this pyramid—a rectangular prism that has three times the volume of the pyramid.

Make the pattern for your prism from the grid at the bottom of the sheet. You can check your solution using rice [or sand] or any other method you choose.

Many students will have discovered the 3-to-1 relationship between prisms and pyramids but not yet understand under what conditions the relationship occurs. This task will give you the opportunity to see which students understand the significance of base and height measurements in this comparison. When students have built their rectangular prisms, ask them to write about how they decided what size to make their prisms.

❖ **Tip for the Linguistically Diverse Classroom** Students who are not writing comfortably in English may draw the steps of their decision-making process, or meet with you individually to describe orally what they did.

Some students may recognize that if they make a prism with the same base and height as the pyramid (2 by 6 by 5 cm), it will have three times the volume. They can determine the approximate height (5 cm) by setting the pyramid next to a ruler held vertically. If they use the length of an edge for the height (6 cm), their prism will have *more* than three times the volume.

Other students will devise methods to make a prism with a different base which still has triple the volume as the pyramid. For example, they may start with a prism whose sides are too high, fill it with three volumes of the pyramid, then cut off the extra length on the sides. This is also a valid method.

When students have finished, ask them to bring their prisms for a class discussion. Students will probably have discovered a variety of prisms that have three times the volume of the pyramid.

Did anybody try a method that didn't work? What was it? What methods worked for making a prism with three times the volume of the pyramid?

After students have discussed their methods, ask them to compare the different prisms they made.

Will the different prisms we built have the same volume or different volumes?

It may not be clear to everyone that all prisms with three times the volume of the pyramid will have equal volume, even if they look different. If needed, test the different prisms with students by directly comparing their volumes with rice or sand.

Sessions 2 and 3 Follow-Up

Homework

Give each student one copy of Student Sheet 12, Puzzle Cube Pattern, and three copies of Student Sheet 13, Puzzle Pyramid Pattern.

This is a puzzle to do at home. First you make the solids—one open cube, and three pyramids. Then you need to figure out how to put the three pyramids together so that they make a cube that fits into the cube box.

Students will need scissors and tape at home. If necessary, they can make their solids at school to take home. Encourage students to share this puzzle with their families.

Note: This puzzle offers a good visual demonstration of the 3-to-1 volume relationship between prisms and pyramids with the same height and base. However, while a rectangular prism can always be decomposed into three pyramids with the same volume, in general these three pyramids will not be identical. The cube is a special case.

Geometric Solids and Their Parts

A geometric *solid* is a shape that has three dimensions—length, width, and height. In mathematics, unlike everyday conversation, these geometric shapes are called "solids" whether they are filled or hollow. There are many types of geometric solids. Some, including spheres, cones, and cylinders, have some curved surfaces. Others, called *polyhedra,* have only flat surfaces. Two common types of *polyhedra* are prisms and pyramids.

Prisms The solids we call prisms have two parallel, congruent faces that are connected by rectangular regions. Either congruent face can be called the base. For a rectangular prism, the choice of base is arbitrary but is generally the face that the prism is resting on. Bases are shaded in the diagram.

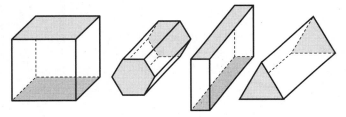

To define prisms in their own words, students might say that "prisms have a top and bottom that are the same shape, and sides that are all rectangles." Of course, this description is accurate only when the prism is turned so that its bases are on the top and bottom.

Pyramids A pyramid consists of one polygonal face (the base) and a point not in the plane of this polygon, which we will call its "top." The top is joined by line segments to each vertex of the polygon, forming lateral faces that are triangular regions. Students might describe a pyramid as having "a flat bottom and a pointed top, with sides that are all triangles."

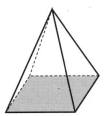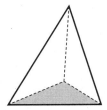

Cylinders Like prisms, cylinders have two parallel and congruent faces (either of which can be referred to as the base). But instead of polygons for a base, they have circles.

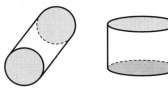

Cones A cone consists of a circular region (the base) which is joined to a point not in the plane of the circle.

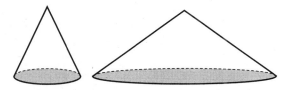

The Height and Diameter The *height* of a pyramid or cone is the length of a perpendicular line segment drawn from the top to the base. The height of a prism or cylinder is the distance from base to base. The *diameter* of a cone or cylinder is the length of a segment that has its endpoints on the circular base and passes through the center of the base.

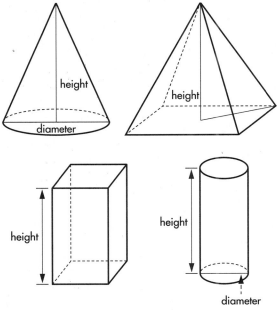

Continued on next page

Some students may use the length of an edge or the side of a pyramid or cone as the height. This is incorrect, but will not be far off for tall, narrow solids. As such figures become flatter (see diagram), the inaccuracy of using the wrong dimension for the height will increase.

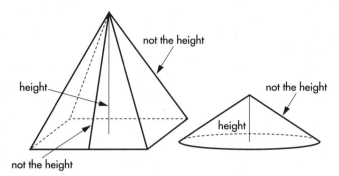

Encourage students to judge the height of pyramids and cones by holding a ruler next to them, as shown on the Dimensions of Solids transparency (p. 147).

Other Terminology   As students talk about solids, they will use various words to describe their parts, and even the figures themselves. For example, in talking about a rectangular prism, they may refer to *corners* instead of *vertices* and *sides* instead of *faces,* and may call the figure a *box* instead of a *rectangular prism.* You can use these everyday terms yourself, but also introduce and encourage the use of the standard terms as they fit into the conversation.

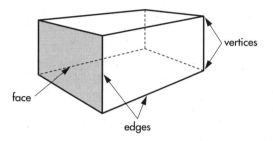

It is not necessary for students to use the standard terms consistently, but they do need to find words that communicate their meanings clearly.

Exploring the 3-to-1 Relationship

This class has been working with the paper solids shown on Student Sheet 10, Pairs of Solids. After the students have compared the volume of the solids in each pair with rice, the teacher places a set of paper solids on a table. She has written the answers that students have agreed upon in a large chart on the board.

Pair	Solids	Number of smaller solid needed to fill larger solid
1	A,B	3
2	C,D	3
3	E,F	3
4	G,H	2
5	I,J	6
6	I,K	3

As you worked with these pairs of solids, what did you find?

Zach: One of the two in each pair has a point.

Rachel: And the other ones are flat tops.

Amir: The smaller one ends in a point.

Becky: The bigger ones are bigger cause they go up *[motioning along two vertical parallel lines with her hands],* and the smaller ones come together *[making a point with her hands].*

Antonio: It looks like in pairs 1, 2, 3, and 6, the bigger one had 3 times the rice as the smaller one. Pairs 4 and 5 were different.

What was special about the solids that have the 3-to-1 relationship?

Kevin: In pair 4, the point [the top of the cone] is higher. In the others [other pairs] the points are even; they're the same.

Cara: The heights are the same, except pair 4.

Manuel: They all fit inside each other, so they're even in height.

Becky: When the heights were the same, the straight ones were three times bigger.

Some of you have talked about the heights of these solids. What does height mean?

[Cara moves to the table and places her hand horizontally across the tops of both solids in each of the first three pairs and pair 6.]

Cara: They're the same height here; they're the same height here; and here, and here.

What do we define as height on these shapes— from where to where?

Sofia: The base *[she holds out a horizontal hand]* to the top of the object *[moves her horizontal hand upwards].*

OK, you said the heights are the same for the solids that are 3-to-1. Is that the only thing that's the same?

Mei-Ling: If you look at the bottoms, if you put them together *[motions with her hands],* they are the same shape.

Are you saying that the bottom of the square prism and square pyramid *[points to pair 1]* are exactly the same size?

[The teacher first holds the two solids with their square bases facing the students; then places the prism on top of the upside-down pyramid so their bases fit together.]

Jasmine: Yeah, the bottoms are the same.

What do we call the bottoms of these figures?

Several students: The base.

What about pair 2 *[showing the bases]*? Are the bases of these the same?

Students: Yes.

How would we prove that the base has something to do with the 3-to-1 relationship?

Desiree: Look, the bases are the same for I and K, but J is different. And the heights are the same. Shape I didn't have 3 times the rice as J.

So is the same height enough?

Leon: No, they have to have the same base.

Using Standard Volume Units

Materials

- Students' solids A–K (from Sessions 2–3)
- Rice or sand, trays
- See-through prisms (1 per small group)
- Centimeter cubes (available)
- Centimeter rulers (1 per pair)
- Liter measuring pitchers (4 per class)
- Calculators (available)

What Happens

With the help of a special measuring tool—a see-through graduated prism—students figure out a method for determining the volume, in cubic centimeters, of each of their 11 solids. Student work focuses on:

- using standard units of volume to measure nonrectangular solids

Activity

Measuring with Cubic Centimeters

If you don't have plastic see-through prisms, copy the pattern (p. 146) on transparency film and assemble enough for your class.

Introduce the see-through prism without telling students that they will use it as a measuring device for the rest of the unit. (Since there are no numbers on the sides, only centimeter squares, its use as a measuring tool will not be obvious.) Explain that in this activity, students will be finding the volume of this prism as well as the volumes of solids A–K.

In the last sessions, you compared the volumes of your different solids. Now your task is to find the actual volumes of these solids as measured in cubic centimeters.

You'll start with the see-through prism. *[Show some centimeter cubes.]* **How many of these cubic centimeters will it take to completely fill this prism? After you figure that out, you'll find the volume of Solid A, the rectangular prism.**

Then you'll try Solid B, the rectangular pyramid *[hold it up]*. **Of course, we can't *really* fill this pyramid with the centimeter cubes—we can't get cubes to fit into the pointed top. But what if these centimeter cubes were made out of clay? Then we could squish them to fit into the pointed top, or any other small space. Do you think you could find out how many "clay" cubic centimeters it would take to fill the rectangular pyramid?**

Today and tomorrow, you need to find and record the volume of all eleven solids you made, measured in cubic centimeters.

Students are to keep their own records on notebook paper, and later enter their results in the class chart for comparison.

You can use rice [or sand], centimeter cubes, your rulers, or whatever other materials you need to solve these problems. Use notebook paper to keep track of any computation you do, because we'll be talking later about how you got your answers.

Make a table on the board for students to record the volume they determined for each solid.

Solid	Volume (cubic centimeters)
see-through prism	
A	
B	
C	
D	
E	

Solid	Volume (cubic centimeters)
F	
G	
H	
I	
J	
K	

To find the number of cubic centimeters in the see-through prism and rectangular prism A, most students will use the methods they developed in Investigations 1 and 2, using the dimensions of the figures. However, some may still need to place actual centimeter cubes in the prisms, at least enough for the bottom layer.

Students will use a variety of methods to determine the volumes of the other solids. For example, they might approach rectangular pyramid B in one of the following ways:

■ They might first find the volume of rectangular prism A by multiplying its length, width, and height. Recalling that the volume of rectangular pyramid B is ⅓ of that, they divide their answer for the prism by 3 to get the volume of the pyramid.

■ They might fill the pyramid with rice and pour that rice into the see-through prism, then determine how many cubic centimeters the rice has filled.

■ They might fill the see-through prism with rice, pour rice from the prism into the pyramid to fill it, and calculate the volume of the unfilled portion of the prism.

Some students may propose to make a cubic centimeter box out of graph paper and use it to fill their solids with rice, one cubic centimeter at a time. This method will work, and shows an excellent understanding of the principles involved; however, it is impractical because of the large number of cubic centimeters that fit in the solids. Be supportive of students who think this way, but suggest that they try to find a more efficient method.

Some students may not think of using the see-through prism as a measuring device. You might suggest the idea with a question:

Can you think of a way this prism could help you find how many cubic centimeters fit in solid B?

If that doesn't help, you can be more directive, modeling its use.

Let's fill pyramid B with rice, and pour this rice into the see-through prism. How many cubic centimeters will it take to fill the see-through prism to the same level as the rice? Knowing that, how many cubic centimeters will it take to fill pyramid B?

While you are observing students, pick out two or three pairs who are working successfully to share their methods with the class in the follow-up discussion.

Note: Plastic see-through prisms, if you have them, will give more accurate measurements than prisms made from transparency film. Because transparency film is not rigid, the sides of the handmade prism will bulge slightly as the rice is added, distorting the measurement. Students can minimize this distortion by pushing in the sides of the prism with their fingers, to keep the sides flat, especially near the top of the prism.

Discussion: How We Found the Volume

After most of the pairs have measured the volume of the see-through prism and all 11 solids, proceed to the following discussion.

What methods did you use for finding the volumes of these solids? Why do you think these methods work? Did anybody try any methods that didn't work?

Direct attention to the class chart where student pairs have recorded the volume they found for each solid.

Let's look at the answers for each solid. Should we all get the same answers?

Students' determination of a solid's volume with rice often differs from the computed volume. This can happen because of inaccuracies in measuring the dimensions of the solids, because of the difficulty of constructing precise solids from the patterns, and because sides may bulge out when rice is poured into solids. For rectangular prisms, the volume measured with rice rarely matches exactly the volume found by multiplying the length, width, and height. Loosely taped or bulging sides can cause surprisingly large discrepancies. Thus, it is important to evaluate students' *methods,* rather than their answers, to assess their understanding of the concepts.

If there are some answers that are *very* different from the rest, ask what methods were used to find these answers. What computations were performed? What were the answers to these computations?

Cubic Centimeters in a Liter

Show students the liter pitcher, which they will be familiar with if you have done the grade 5 unit, *Measurement Benchmarks*.

This pitcher holds 1 liter. A liter is one way to measure liquid volume. What do you estimate the volume of this pitcher is in cubic centimeters? Each of you write your estimate on a slip of paper. Then, in small groups, figure out the actual answer.

Distribute the measuring pitchers. Students can use their see-through prisms and rice or sand to find the answer.

Collect each group's answer and discuss any discrepancies (which may arise from the bulging sides if you are using prisms made from transparency film). Lead students to decide on a value for the volume of the pitcher. One liter has a volume of exactly 1000 cubic centimeters, so five see-through prisms, each 200 cubic centimeters, should fill the liter.

Who remembers how many milliliters there are in a liter? So, how does a milliliter compare to a cubic centimeter? (1 milliliter is one thousandth of a liter, so it is equal to 1 cubic centimeter.)

Once students understand the equivalence between cubic centimeters and milliliters, they can use the liter pitcher for measuring the volume of large containers. This may be useful for measurements students will be doing as part of their final project.

How Do the Heights Compare?

What Happens

Students determine what height to make a cylinder so that it has the same base and volume as a given cone. Then they figure out the dimensions of a square prism that will have the same base and volume as a given square pyramid. Student work focuses on:

■ exploring volume relationships between related solids

Materials

■ Rice or sand, trays

■ Centimeter rulers (1 per pair)

■ Scissors and tape for each pair

■ Cone, Cylinder, and Pyramid Patterns (1 per pair)

■ One-centimeter graph paper (1 per pair, plus extras)

■ Calculators (available)

Activity

Comparing Cylinders and Cones

In Sessions 2 and 3, the students worked with a cylinder and cone that had the same base and height (cylinder C and cone D), and they determined how the volumes of two such solids were related. In this excursion, they explore the relationship from another perspective: Here they are considering a cylinder and a cone that have the same volume and base, and their task is to determine how the heights are related.

Give each student pair a copy of the cone pattern and the cylinder pattern. Using these patterns, students are to design a cylinder that has the same volume and base as the cone. That is, they must determine the height the cylinder needs to be in order to have the same volume as the cone.

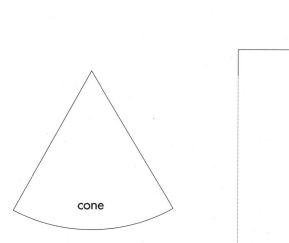

If the base and volume of this cone and cylinder are the same, how will their heights compare?

Some students may approach this problem by guessing at the height for the cylinder, then testing their guess by pouring rice from the cone into the cylinder. If their guess gives them a cylinder that's too tall, they can simply cut it down to find a solution. But if it is too short to start with—or if they cut it too short in making an adjustment—they might then have difficulty figuring out how much taller it needs to be. Therefore, you might encourage students to *mark* the proposed height but not actually cut the cylinder down until they check their guess.

Other students might approach the problem through computation, considering the 3-to-1 volume relationship (between cylinders and cones with the same height and base) that they discovered earlier in this investigation. In this case, since the volume is to be the same, they would figure the height of the cylinder to be one-third the height of the cone.

When they have solved this problem, students compare their cylinders and their methods with other pairs working near them. If you would like the class to hear about other methods, ask for their attention briefly and call on two or three students with different successful methods to describe them.

Activity

Comparing Prisms and Pyramids

Give each student pair a copy of the pyramid pattern and a sheet of one-centimeter graph paper.

This time, we want to end up with a square prism that has the same volume and base as this pyramid. If the base and volume of the prism and pyramid are the same, how will their heights compare? What will be the dimensions of your square prism? Construct your square prism from the graph paper.

By now, most students should be comfortable making patterns for rectangular prisms. Suggest that they try making their prism patterns with just one sheet of paper, if possible.

Building Models

What Happens

As a final project, students design and create a model made from geometric solids. The model must include prisms, pyramids, cylinders, and cones, all of which students can create from patterns. Students describe each solid they use in terms of its dimensions and volume. They estimate and then determine the actual total volume of their model in cubic centimeters. Student work focuses on:

- using geometric solids to make real-world objects
- determining the volume of geometric solids
- describing methods for determining volumes of geometric solids

Materials

- Student Sheet 14 (1 per small group)
- Student Sheet 15 (1 per small group)
- Dimensions of Solids transparency
- Pattern Makers (1 of each per small group, plus extras)
- One-centimeter graph paper (1 per small group, plus extras)
- Scissors and tape for each pair
- See-through prisms, liter pitchers, centimeter cubes, centimeter rulers (available)
- Rice or sand, trays
- Calculators (available)
- Overhead projector

A Final Project with Geometric Solids

Students work in groups of two to four on the final project of this unit: designing a three-dimensional model made out of paper prisms, pyramids, cylinders, and cones. This work encourages students to see geometric solids as parts of real-world objects. For example, they might use a flat cone to represent a headlight on a car; cylinders might represent rocket engines; various geometric solids can serve as robot body parts. As students plan their models, they imagine how different solids look as they are rotated to various positions, which gives them good practice in visualization.

Students can be extremely creative on this project. Actual models have included robots and racing cars; a jungle gym for hamsters; a hovercraft powered by solar energy; a castle playground; a home for small animals; a stick-on radio that attached to the listener's shoulder with Velcro pyramids; a flying bug shaped like an airplane; and an environmental elevator, with a passenger compartment that really went up and down.

You need to set a few limits to keep the project manageable. Students must use at least 8 and no more than 10 solids in their model, including at least one pyramid, one prism, one cone, and one cylinder.

Distribute Student Sheet 14, Final Project Tasks, and Student Sheet 15, Model Planning Sheet, to each small group. Go over these sheets with the class as you introduce the project. Explain that part of their job as model designers will be to determine the volume of each solid they use, as well as the total volume of their model, in cubic centimeters. This information is needed so a company could determine how much material would be required to produce their design. First they will estimate what the total volume of their model will be; then they will determine its actual volume by adding the actual volumes of the individual solids.

❖ **Tip for the Linguistically Diverse Classroom**   Model each direction on the task sheet to ensure comprehension before students begin the tasks.

Give each small group a packet of Pattern Makers, which provide patterns for cones, cylinders, rectangular and square pyramids, and triangular prisms of different sizes. Allow some time for them to look through the patterns and to ask questions about their use. Also distribute one-centimeter graph paper for making rectangular prisms.

Task 1: Planning the Models The first task for each group is to decide what their model will be. With pencil and paper, they prepare a sketch. On the Model Planning Sheet, they write a brief description. For example:

> It's a space station that has living quarters and two booster rockets on the outside.

Working with their sketch and the patterns available, students plan the size of each solid. To keep track of their work, students should label each solid in their model with an identification number. They record the dimensions of each numbered solid in the chart on their Model Planning Sheet.

Students will be familiar with specifying the three dimensions of a prism. However, although they have talked previously about the height of pyramids, cones, and cylinders, they may be unfamiliar with other dimensions for these solids. To illustrate how to record these dimensions, show the transparency Dimensions of Solids (p. 147) on the overhead. Remind students that they can find the height of a cone or pyramid by placing it on its base and putting a ruler next to it, perpendicular to the base. Also, explain how we measure the diameter of a circular base. Check to see that students understand which dimensions they are to list on the Model Planning Sheet.

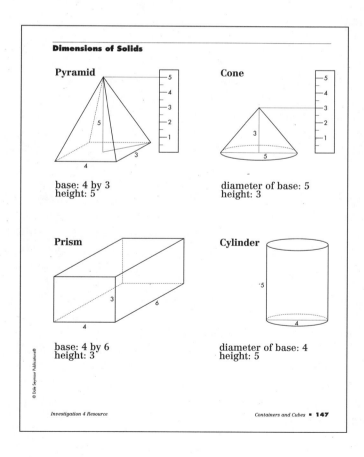

Task 2: Making the Solids The second task for students is to make a pattern for each solid in their model. Show them where they can get additional copies of centimeter graph paper and Pattern Maker sheets as needed, along with scissors and tape.

Directions for use of the Pattern Makers are given on each sheet. Read through these instructions with students as needed. The size and shape of the solid is determined by how the pattern is cut out. Paths on the patterns are labeled with the dimensions of the solid that each path will produce. Students should need no help in designing rectangular prisms (boxes) on the centimeter graph paper.

Students cut out, fold, and tape their patterns to make the solids (just as they did with solids A–K earlier in the investigation).

Task 3: Predicting the Volume of the Model After students have put together the individual solids that make up their model, their third task is to predict the total volume of their model. They write this prediction on a sheet of notebook paper, and describe in writing how they came up with their prediction.

❖ **Tip for the Linguistically Diverse Classroom** Students who are not writing comfortably in English can use simple sketches to demonstrate their method of predicting.

Task 4: Finding the Actual Volume Next, students determine the volume of each solid in cubic centimeters (rounded to the nearest whole cubic centimeter). They record these volumes in the chart on their Model Planning Sheet, and they describe on the back how they found the volume of each solid. All the students in the group should take turns doing this writing for different solids.

While students are finding the volumes of their solids, circulate and ask groups to explain their methods. If they are using rice for rectangular prisms and pyramids, ask them if they can find other ways to do this. For instance, encourage students to compute the volume of these prisms by using the methods they developed at the beginning of the unit. Students also could find the volume of a rectangular pyramid by computing the volume of a rectangular prism with the same base and height, then dividing by 3 (but don't push this method too hard; some students might not be completely comfortable with it). Some students might even correctly find the volume of the triangular prism by taking half the volume of a rectangular prism that has the same dimensions.

See the **Dialogue Box,** Finding the Volume of Our Models (p. 104), for some ideas about what to expect as you circulate during this part of the activity.

Task 5: Building the Models   Once they have found the volume of each solid, students tape or glue them together to build their models. Some will also want to draw on or color parts of their models.

Task 6: The Final Report   The last task for students is to compile a final report. This consists of their sketch of the model, the actual model, a description of how they estimated the total volume of their model, and their completed Model Planning Sheet.

Presenting Models to the Class

When the groups have completed their models and compiled their final reports, they are ready to present their findings. Before the presentations start, students create a display in which they attempt to put their models in order of increasing volume, using quick visual estimation. Make a note of any disagreements to check later when students report the computed volume of their models.

Explain the presentation format: Groups will take turns standing before the class and explaining the following:

1. What their model is
2. The volume they predicted for their model
3. How they found the volumes of the solids that make up their model
4. The total volume of the model

You might list these items on chart paper or the board so that students keep them in mind as they are presenting.

One responsibility of the listeners, or "review board," is to evaluate the validity of the methods each group used for finding volume and the reasonableness of their results.

❖   **Tip for the Linguistically Diverse Classroom**   Be sure groups understand that all students are to take an active role in these presentations. Those with limited English proficiency may hold up visuals, point to objects or parts of the model, and demonstrate the groups' methods while English-proficient students speak.

Before a group reveals the actual volume of any part of the model, the remaining students estimate the model's total volume, noting their estimate on a slip of paper. Regularly ask students how they made their estimates. Estimating the volume of each model will help keep the students' focus on volume (it is like playing a game; students really enjoy this). It also keeps the whole class involved as the group presentations are given.

Allow students to use calculators in making their volume estimates. For example, students will typically estimate the volume of a rectangular prism by estimating the dimensions and then multiplying, reinforcing the methods used for finding volume earlier in the unit.

If a group reports that they used rice for determining the volume of all their solids, ask if there is another way to find volume for some of them—the prisms and pyramids in particular.

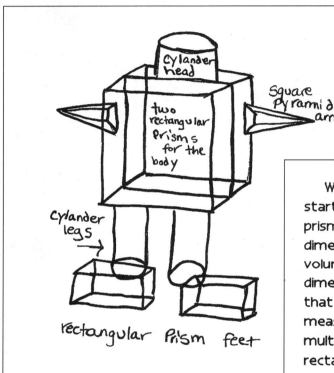

When we had to find the volume of the robot we started with the body because it was a rectangular prism on centimeter paper and we could count the dimensions. We multiplied the dimensions and found the volume. With the other shapes we couldn't count the dimensions, so we filled the shapes and then poured that into the see-through prism with the centimeters measured on it. We counted those dimensions and multiplied them together the same way we did for the rectangular prism.

Our results were the following: the dimensions of the body were 10X16X4 = 640 cubic cm. X 2 = 1,280. The volume of the legs were 304X2 = 608. The volume of the arms was 64X2 = 128. The volume of the robot's feet was 160X2 = 320 and the volume of the head was 320. We multiplied the legs, arms and feet by 2 because there are two of them. We doubled the body because we used two sheets of paper and counted the body as two solids and we measured one of the shapes. Our total volume was 2,656.

Collect the students' models and their final reports to assess their understanding of the concepts presented in this unit. Consider the following as you review student work:

■ Can students determine the volume of a solid in cubic centimeters?

■ What strategies are students using to find the volume of prisms? As needed, refer back to the **Teacher Note,** Strategies for Finding the Number of Cubes in 3-D Arrays (p. 26), for a discussion of strategies of varying degrees of sophistication.

■ Do students use any volume relationships in their figuring? For example, do they use the idea that the volume of a pyramid is $\frac{1}{3}$ the volume of a prism with the same base and height?

In assessing students' work, recall that students might use correct methods but still get inaccurate answers because of small measurement or construction errors. Be sure to evaluate the validity of students' *methods* rather than the accuracy of their answers.

As the unit ends, encourage students to spend 20–30 minutes revisiting their work for the last few weeks. You may want to use one of the following options for creating a record of students' work on this unit:

■ Students look back through their folders or notebooks and write about what they learned in the unit, what they remember most, and what was hard or easy for them. You might have students complete this work during their writing time.

■ Students select one or two pieces of their work as their best, and you also choose one or two pieces of their work to be saved. This work is saved in a portfolio for the year. You might include students' written responses to checkpoint and assessment activities, including their report for the final project. Students can create a separate page with brief comments describing each piece of work.

■ You may want to send a selection of work home for families to see. Students write cover letters, describing their work in this unit. This work should be returned if you are keeping a year-long portfolio of mathematics work for each student.

Sessions 7, 8, and 9 Follow-Up

Extension

Doubling Amounts of Rice   Read aloud *A Grain of Rice,* by Elena Pittman (Hastings House, 1986). In the story, the emperor's daughter falls in love with a peasant boy, Pong Lo. Even though Pong Lo later saves the daughter's life, they are forbidden to marry because he is not noble and rich. When asked by the emperor what prize he wishes for saving the daughter's life, Pong Lo says 1 grain of rice on the first day, 2 on the second, 4 on the third, and keep doubling the amount for 100 days. As the amount of rice increases, the author describes how the containers required to give Pong Lo his daily due increase in size. After 40–45 days, the emperor is having great difficulty finding rice, and Pong Lo has become rich by selling rice. The marriage is allowed, and all live happily ever after.

Students enjoy this story, marveling at the enormity of the numbers involved. The connection between the amount of rice and its volume is not made as clear as it could be for our purposes, but you can still make this connection an interesting extension. The table below illustrates this idea.

In *A Grain of Rice,* how much rice would Pong Lo get, according to his plan?			
Day	Grains of rice		
1	1	1	
2	2	2	
3	$2 \times 2 = 2^2$	4	
4	$2 \times 2 \times 2 = 2^3$	8	
5	$2 \times 2 \times 2 \times 2 = 2^4$	16	
6	2^5	32	
7	2^6	64	
8	2^7	128	
9	2^8	256	
10	2^9	512	
20	2^{19}	524,288	
30	2^{29}	536,870,912	$\approx$ 536 million
40	2^{39}	$\approx$ 549,755,813,900	$\approx$ 549 billion
50	2^{49}	$\approx$ 562,949,953,400,000	$\approx$ 562 trillion
60	2^{59}	$\approx$ 576,460,752,300,000,000	$\approx$ 576 quadrillion
70	2^{69}	$\approx$ 590,295,810,400,000,000,000	$\approx$ 590 quintillion
80	2^{79}	$\approx$ 604,462,909,800,000,000,000,000	$\approx$ 604 sextillion
90	2^{89}	$\approx$ 618,970,019,600,000,000,000,000,000	$\approx$ 618 septillion
100	2^{99}	$\approx$ 633,825,300,100,000,000,000,000,000,000	$\approx$ 633 octillion
Total		$\approx$ 1,267,650,600,000,000,000,000,000,000,000	$\approx$ 1.267 nonillion

How much space would be taken up by Pong Lo's rice?

- Let's assume that one cubic centimeter contains about 20 grains of rice. A cubic meter contains 1 million cubic centimeters, so 1 cubic meter would contain 20 million (20,000,000) grains of rice.

- Suppose the volume of a classroom is 288 cubic meters. The amount of rice that fits in the classroom would be 288×20 million = 5,760,000,000 = 5.76 billion grains. This is about how much rice was given to Pong Lo on the 33rd day.

- It would take about 95 classrooms to hold the rice that Pong Lo was given on the 40th day, and 97,734 classrooms to hold the rice that he should be given on the 50th day.

- It would take about 220,000,000,000,000,000,000 classrooms to hold the *total* amount of rice that Pong Lo would be given in the 100 days. This amount of rice would fill about 56 spheres the size of the earth.

━ D I A L O G U E ▢ B O X ━

Finding the Volume of Our Models

While small groups were working together on the final project for this unit, the teacher circulated to observe their methods for finding the volume of the solids in their models.

How did you find the volume for this pyramid?

Jeff: Side 1 × side 2 × height, divided by 3.

Why did you divide by 3?

Jeff: Because there are three numbers, side 1, side 2, and the height.

[Speaking to Jeff's partners, Antonio and Yu-Wei] **What do you two think?**

Antonio: If you take a prism just like it, the same base and height, it would be one-third its volume.

Yu-Wei: If you multiply length, width, height, you get the prism. If it's the same base and height as the pyramid, then it's 3 times as much. So divide by 3 [for the prism].

[The teacher notices that the three boys have calculated the volume of a cylinder by multiplying the diameter by the height.]

So 14 cubic centimeters fit in the cylinder?

Antonio: No, I don't think that's right. [Jeff agrees.]

How did you find the volumes of cylinders and cones before?

Antonio: We used rice, and poured it into the prism with squares marked on it.

[The boys decide to use this same procedure here. The rice from the cylinder fills the see-through prism 8 layers high.]

Jeff: So, 10 × 8 = 80.

Why multiply by 10?

Jeff: Because there's 10 in the bottom.

[The teacher then notices that the boys have

written 8 for the volume of a cone. Thinking that this measurement is a bit off, the teacher shows the boys a plastic centimeter cube.]

If I melted 8 of these down and poured them into the cone, would they completely fill it?

[All three of the boys say no, and then use rice to estimate that there are really 15 cubic centimeters in the cone.]

[Continuing to circulate, the teacher discovers that three girls have found a novel way to determine the volume of very small solids.]

How did you find the volume of this little cone?

Shakita: We filled it with rice, then poured the rice into the see-through prism. We tilted the prism so that the rice went into a corner. It looked like about 2 cubic centimeters of rice *[which was a very a close estimate]*.

Julie: I made a cubic-centimeter box and used three of them to fill this other little cone.

Amy Lynn: We did this small cylinder another way. We kept filling the see-through prism with cylinders of rice until it got to 1 [centimeter high]. Since it took 5 cylinders, and 10 cubes fit in there [the bottom layer of the prism], you divide 10 by 5 and get 2 cubic centimeters for the cylinder.

Counting Around the Class

Basic Activity

Students count around the class by a particular number. That is, if counting by 2's, the first student says "2," the next student says "4," the next "6," and so forth. Before the count starts, students try to predict on what number the count will end. During and after the count, students discuss relationships between the chosen factor and its multiples.

Counting Around the Class is designed to give students practice with counting by many different numbers and to foster numerical reasoning about the relationships among factors and their multiples. Students focus on:

- becoming familiar with multiplication patterns
- relating factors to their multiples
- developing number sense about multiplication and division relationships

Materials

Calculators (for variation)

Procedure

Step 1. Choose a number to count by. For example, if the class has been working with quarters recently, you might want to count by 25's.

Step 2. Ask students to predict the target number. "If we count by 25's around the class, what number will we end up on?" Encourage students to talk about how they could figure this out without doing the actual counting.

Step 3. Count around the class by your chosen number. "25...50...75..." If some students seem uncertain about what comes next, you might put the numbers on the board as they count; seeing the visual patterns can help students with the spoken one.

You might count around a second time by the same number, starting with a different person, so that students will hear the pattern more than once and have their turns at different points in the sequence.

Step 4. Pause in the middle of the count to look back. "We're up to 375 counting by 25's. How many students have we counted so far? How do you know?"

Step 5. Extend the problem. Ask questions like these: "Which of your predictions were reasonable? Which were possible? Which were impossible?" (A student might remark, for example, "You couldn't have 510 for 25's because 25 only lands on the 25's, the 50's, the 75's, and the 100's.")

"What if we had 32 students in this class instead of 28? Then where would we end up?"

"What if a different number of students counted? This time we counted by 25's and ended on 700; what if we counted by 50's? What number do you think we would end on? Why do you think it will be twice as big? How did you figure that out?"

Variations

Multiplication Practice Use single-digit numbers to provide practice with multiplication (that is, count by 2's, 3's, 4's, 5's, 7's, and so forth). In counting by numbers other than 1, students usually first become comfortable with 2's, 5's, and 10's, which have very regular patterns. Soon they can begin to count by more difficult single-digit numbers: 3, 4, 6, and (later) 7, 8, and 9.

Landmark Numbers When students are learning about money or about our base ten system of numeration, they can count by 20's, 25's, 50's, 100's, and 1000's. Counting by multiples of 10 and 100 (30's, 40's, 600's) will support students' growing familiarity with the base ten system of numeration.

Making Connections When you choose harder numbers, pick those that are related in some way to numbers students are very familiar with. For example, once students are comfortable counting by 25's, have them count by 75's. Ask students how knowing the 25's will help them count by 75's. If students are fluent with 3's, try counting by 6's or by 30's. If students are fluent with 10's and 20's, start working on 15's. If they are comfortable counting by 15's, ask them to count by 150's or 1500's.

Large Numbers Introduce large numbers, such as 2000, 5000, 1500, or 10,000, so that students begin to work with combinations of these less familiar numbers.

What Could We Count By? Pick a number such as 100. Ask students what number they could count by to be sure that someone in the class would say 100. Encourage children to share their strategies for figuring this out. Count around the class by the suggested numbers to see if they work. Repeat with other numbers. That is, what could they count by to be sure that someone in the class would say 50? 1000? 24?

Don't Start with 0 Continue to count around the classroom, but start the count with a multiple other than 0. For instance, you might count by 10's or 25's, but start at 50, 100, 1000, or 525.

Fractions and Decimals Count around the classroom by fractions or by decimal numbers. Begin with easier landmark numbers such as 0.5, 0.25, one-third, and three-fourths. As you feel your students are ready, count by more difficult numbers (such as ½, 2.25, and 0.125), or count by messier decimals on the calculator, and discuss any patterns your students see.

Counting Backwards Starting with a given number, count backwards around the class by numbers whose patterns are familiar to students (such as 2's, 5's, 10's, and 25's). As students become more comfortable with this variation, try counting by more difficult numbers. Or, play a modified version of What Could We Count By? Give students a starting number (such as 100 or 1000) and ask them to find numbers they can count backwards by that will land them exactly on 0 (or so that someone will say a particular number somewhere in the count).

Using the Calculator On some days everyone or a few students might use calculators to skip count while you are counting around the class. On most calculators, the equals (=) key provides a built-in constant function, allowing you to skip count easily. For example, if you want to skip

count by 25's, you press your starting number (let's say 0), the operation you want to use (in this case, +), and the number you want to count by (in this case, 25). Then, press the equals key each time you want to add 25. So, if you press

you will see on your screen 25, 50, 75, 100, 125.

Special Notes

Letting Students Prepare When introducing an unfamiliar number to count with, students may need some preparation before they try to count around the class. Ask students to work in pairs, with whatever materials they want, to determine on what number the count will end.

Avoiding Competition Be sensitive to potential embarrassment or competition that may occur if some students have difficulty figuring out their number. One teacher allowed students to volunteer for the next number, rather than counting in a particular order. Other teachers have made the count a cooperative effort, establishing an atmosphere in which students readily helped each other, and anyone felt free to ask for help.

Related Homework Options

Counting Patterns Students write out a counting pattern up to a target number (for example, by 25's up to 500). Then they write about what patterns they see in their counting. Calculators can be used for this.

Mystery Number Problems Provide an ending number and ask students to figure out what factor they would have to count by to reach it. For example: "I'm thinking of a mystery number. I figured out that if we counted around the class by my mystery number today, we would get to 2800. What is the mystery number?"

Or, you might provide students with the final number and the factor and ask them to figure out the number of students in the class. "When a certain class counts by 25's, the last student says 550. How many students are in the class?" Calculators can be used.

Guess My Number

Note: The Ten-Minute Math activities suggested for this unit involve Guess My Unit, a variation of the Guess My Number activity.

Basic Activity

You choose a number for students to guess, and start by giving clues about the characteristics of the number. For example: It is less than 50. It is a multiple of 7. One of its digits is 2 more than the other digit.

Students work in pairs to try to identify the number. Record students' suggested solutions on the board and invite them to challenge any solutions they don't agree with. If more than one solution fits the clues, encourage students to ask more questions to narrow the field. They might ask, for example: Is the number less than 40? Is the number a multiple of 5?

Guess My Number involves students in logical reasoning as they apply the clues to choose numbers that fit and to eliminate those that don't. Students also investigate aspects of number theory as they learn to recognize and describe the characteristics of numbers and relationships among numbers. Students' work focuses on:

- systematically eliminating possibilities
- using evidence
- formulating questions to logically eliminate possible solutions
- recognizing relationships among numbers, such as which are multiples or factors of each other
- learning to use mathematical terms that describe numbers
- sorting measuring units

Materials

- 100 chart or 300 chart (optional)
- Scraps of paper or numeral cards for showing solutions (optional)
- Calculators (for variation)
- Guess My Unit cards (pp. 157–158), cut into decks (one per pair or group, for Guess My Unit)

Procedure

Step 1. Choose a number. You may want to write it down so that you don't forget what you picked.

Step 2. Give students clues. Sometimes, you might choose clues so that only one solution is possible. Other times, you might choose clues so that several solutions are possible. Use clues that describe number characteristics and relationships, such as factors, multiples, the number of digits, and odd and even.

Step 3. Students work in pairs to find numbers that fit the clues. A 100 chart (or 300 chart for larger numbers) and scraps of paper or numeral cards are useful for recording numbers they think might fit. Give students just one or two minutes to find numbers they think might work.

Step 4. Record all suggested solutions. To get responses from every student, you may want to ask students to record their solutions on scraps of paper and hold them up on a given signal. Some teachers provide numeral cards that students can hold up to show their solution (for example, they might hold up a 2 and a 1 together to show 21). List on the board all solutions that students propose. Students look over all the proposed solutions and challenge any they think don't fit all the clues. They should give the reasons for their challenges.

Step 5. Invite students to ask further questions. If more than one solution fits all the clues, let students ask yes-or-no questions to try to eliminate some of the possibilities, until only one solution remains. You can erase numbers as students' questions eliminate them (be sure to ask students to tell you which numbers you should erase). Encourage students to ask questions that might eliminate more than one of the proposed solutions.

Variations

New Number Characteristics During the year, vary this game to include mathematical terms that describe numbers or relationships among numbers that have come up in mathematics class. For example, include factors, multiples, doubling (tripling, halving), square numbers, prime numbers, odd and even numbers, less-than and more-than concepts, as well as the number of digits in a number.

Large Numbers   Begin with numbers under 100, but gradually expand the range of numbers that you include in your clues to larger numbers with which your students have been working. For example:

> It is a multiple of 50. It has 3 digits. Two of its digits are the same. It is not a multiple of 100.

Guess My Fraction   Pick a fraction. Tell students whether it is smaller than ½, between ½ and 1, between 1 and 2, or bound by any other familiar numbers. You might use clues like these:

> It is a multiple of ¼ (for example, ½, ¾, 1 whole, 1¼).

> The numerator is 2 (for example, ⅔, ⅖).

> You can make it with pattern blocks (for example, ⅔, ⅚).

Guess My Unit   In this version of the game, students use logical reasoning to guess a particular unit of measurement. Select a measurement unit on the cards or choose one of your own. If you create a new card, be sure to add that card to each set. Each pair or group of students should have a complete set of Guess My Unit cards, displayed faceup on their desks.

Follow the procedure for the basic game. Give students a few beginning clues that focus on the characteristics and relationships among measurement units. For example: It is a measurement of weight. It is a metric unit. It is more than twice as much as a pound.

Record the initial clues where students can refer to them. Give students a few minutes to work together to try to discover your unit. They may flip over or set aside any cards that your clues

have eliminated as possibilities. They may then ask yes-or-no questions until they can identify the unit.

Calculator Guess My Number   Present clues that provide opportunities for computation using a calculator. For example:

> It is larger than 35×20. It is smaller than $1800 \div 2$. One of its factors is 25. None of its digits is 7.

Don't Share Solutions Until the End   As students become more practiced in formulating questions to eliminate possible solutions, you may want to skip step 4. That is, student pairs find all solutions they think are possible, but these are not shared and posted. Rather, in a whole-class discussion, students ask yes-or-no questions, but privately eliminate numbers (or measurement units) on their own list of solutions. When students have no more questions, they volunteer their solutions and explain why they think their answer is correct.

Related Homework Options

Guess My Number Homework   Prepare a sheet with one or two Guess My Number problems for students to work on at home. As part of their work, students should write whether they think only one number fits the clues or whether several numbers fit. If only one solution exists, how do they know it is the only number that fits the clues? If more than one solution is possible, do they think they have them all? How do they know?

Students' Secret Numbers   Each student chooses a number or unit and develops clues to present to the rest of the class. You'll probably want to have students submit their numbers and clues for your review in advance. If the clues are too broad (for example, 50 solutions are possible) or don't work, ask the students to revise their clues. Once you approve the clues, students are in charge of presenting them, running the discussion, and answering all questions about their number during a Ten-Minute Math session.

The following activities will help ensure that this unit is comprehensible to students who are acquiring English as a second language. The suggested approach is based on *The Natural Approach: Language Acquisition in the Classroom* by Stephen D. Krashen and Tracy D. Terrell (Alemany Press, 1983). The intent is for second-language learners to acquire new vocabulary in an active, meaningful context.

Note that *acquiring* a word is different from *learning* a word. Depending on their level of proficiency, students may be able to comprehend a word upon hearing it during an investigation, without being able to say it. Other students may be able to use the word orally, but not read or write it. The goal is to help students naturally acquire targeted vocabulary at their present level of proficiency.

We suggest using these activities just before the related investigations. The activities can also be led by English-proficient students.

Investigation 1

box, layer, stack, predict

1. Show and identify a small, empty *box*.

2. Line up cubes to fill the bottom of the box. As you point to the cubes, explain that you have made one *layer*.

3. Indicate the amount of space left inside the box. Ask students to *predict* how many more layers of cubes will fit in the box.

4. Start a second layer. Explain that you will *stack* the second layer on top of the first. Continue to fill the box, one layer at a time. Keep a tally of layers on the board. When the box is full, students compare the actual number of layers with their prediction.

5. Ask students to name other objects that they could stack in layers to fill the box. Allow them to try their ideas. As they do so, interrupt to point out how many layers thus far.

method

1. Wrap a small box as if for a gift. Show this to the students. Explain that there are different *methods* for unwrapping.

2. Begin by slowly and carefully untaping one side. As you do so, say that this is one *method* for unwrapping a box.

3. Now quickly rip off the paper. Explain that this is another *method* for unwrapping boxes. Point out that both methods work—they unwrap the box.

4. Next, show an orange (with peel). Ask students to show their method for peeling this fruit (that is, where do they begin, how do they get started, and so forth). Point out the differences between methods, emphasizing that there is not one correct way to peel an orange; several methods work.

Investigation 3

space

1. Show students an empty box. Put your hand inside the box as you explain that only air is taking up the *space* inside the box.

2. Show a larger empty box, and place the two boxes side by side. Ask students whether this new box has more or less space inside than the first box.

3. Start to fill one of the boxes with something (such as cubes or balls of crumpled paper). Point out that these items are taking up some of the box's space. Ask:

   Has all the space in the box been filled? Is there more space inside?

4. Supply a variety of small boxes and ask students to place them in order from smallest (least space) to largest (most space).

Investigation 4

model

1. Bring in a variety of models—cars, trains, airplanes, the buildings that go with train sets, and so forth. Explain that these are models—a small version of something that's larger in real life.

2. Ask questions about the models you have brought in.

   Which model is the tallest?
   Which model looks most like the real thing?
   Which model would you like to own?
   What other models could we build?

Blackline Masters

_____, 19_____

Dear Family,

For the next few weeks, our class will be doing a mathematics unit called
Containers and Cubes. The work in this unit—measuring in three dimensions—
has practical uses in daily life. Which bedroom is bigger? How much space do
you have in the trunk of your car? How does the sanitation department measure
your garbage—by weight or volume? How is the size of garbage bags reported?
Knowing how to measure and think spatially helps solve many problems in
mathematics, engineering, and science.

The first problems we will do are called packing problems. Your child will be
finding how many cubes and how many rectangular packages will fit in different
boxes. At home you can talk about similar problems. Many household items are
packaged and sold in boxes. Looking at one layer of these items, can you figure
out how many will fit in the whole box? For example, looking at one graham
cracker, can you figure out how many fit in the whole box? How many cereal,
juice, or toothpaste boxes will fit in a larger rectangular carton, or in your cup-
board space? How many cans of tomato paste will fit in a carton? Rather than
explaining "how to do it," admire the strategies your child develops for figuring
out such problems.

Later in the unit, we will be exploring the properties of geometric solids like
these:

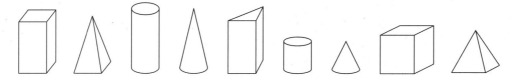

Help your child find examples of these shapes in the real world. Cylinders and
rectangular prisms will be easy to find; the other shapes more challenging. Ask
about the models your child will be building with these shapes. Your interest
and encouragement will be appreciated.

Sincerely,

How Many Cubes? (page 1 of 2)

How many cubes fit in each box? Predict. Then build a box and use cubes to check. Check your prediction before going on to the next box.

Think about a way you could predict the number of cubes that would fit in any box.

	Pattern	Picture	Prediction	Actual
Box 1	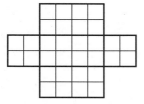	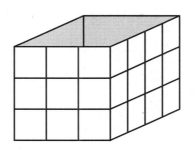	_____	_____
Box 2	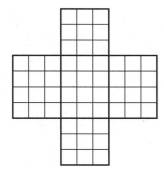		_____	_____
Box 3	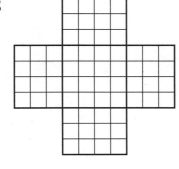	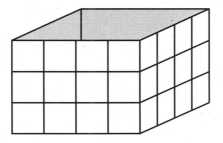	_____	_____

How Many Cubes? (page 2 of 2)

	Pattern	Picture	Prediction	Actual

Box 4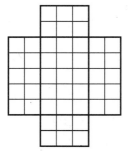

                                                                     _____   _____

Box 5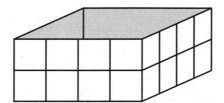

                                                                     _____   _____

Box 6   The bottom of the box is 4 units by 5 units.
The box is 3 units high.   _____   _____

A Method for Predicting

1. Describe a way to predict how many cubes will fit in a rectangular box. Your method should work for any box, whether you start with a box pattern, a picture of the box, or a description of the box in words.

2. Find the number of cubes that fit in a box that is 20 units by 10 units on the bottom, and 12 units high. How can you convince your classmates that your answer is correct?

Predicting Numbers of Cubes

1. Predict how many unit cubes
    are in this package.

    Prediction: _____

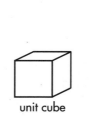

unit cube

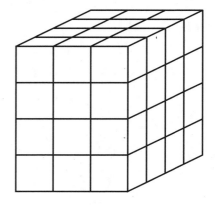

Use graph paper to make a box for this package.
Your box should completely cover all but the top
of the package. The package should completely fill
the box.

Now how many cubes do you predict are in
the package?

    Prediction: _____

2. Predict how many unit cubes
    are in this package.

    Prediction: _____

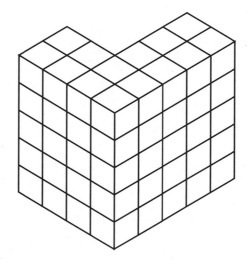

Painting Cubes

Make your predictions by looking at the picture.
Check by making the package with cubes.

1. Find how many unit cubes are in
    this cube package.

    Number of cubes:

    _____

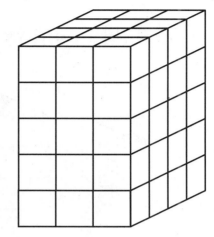

Suppose that the 6 sides of this package were completely
covered with red paint.

2. How many of the cubes would have 0 faces painted? _____

3. How many of the cubes would have 1 face painted? _____

4. How many of the cubes would have 2 faces painted? _____

5. How many of the cubes would have 3 faces painted? _____

6. How many of the cubes would have 4 faces painted? _____

7. How many of the cubes would have 5 faces painted? _____

8. How many of the cubes would have 6 faces painted? _____

Packaging Factory

One ornament is packed in a cube.

Cubes are packed in shipping boxes.

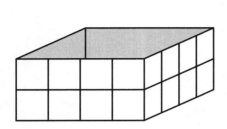 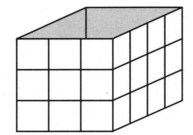

To make a shipping box:

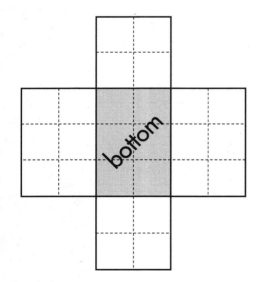 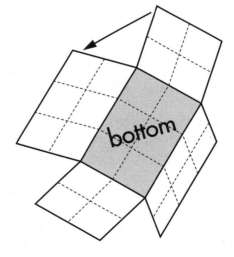 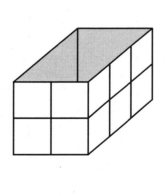

1. Cut out the box pattern.

2. Fold up the sides.

3. Tape edges to make a box.

More Boxes for Predicting

1. How many cubes will fit?

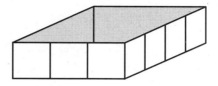

2. How many cubes will fit?

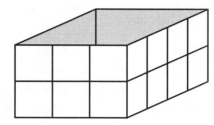

3. How many cubes will fit?

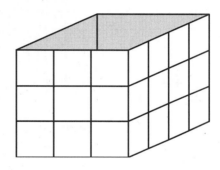

4. How many cubes will fit?

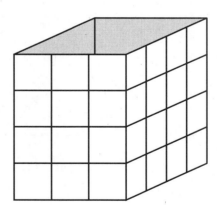

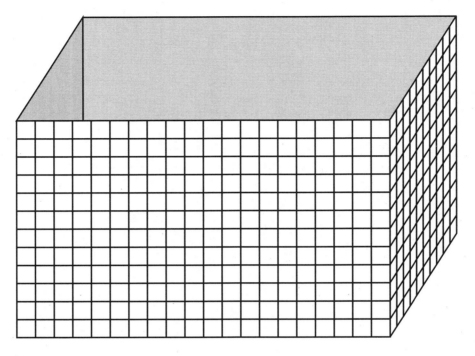

A box 20 by 10 by 12

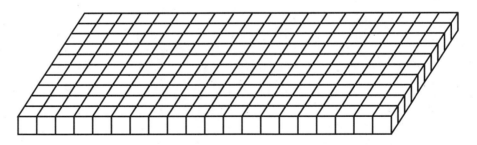

Bottom layer of the box

How Many Packages? (page 1 of 4)

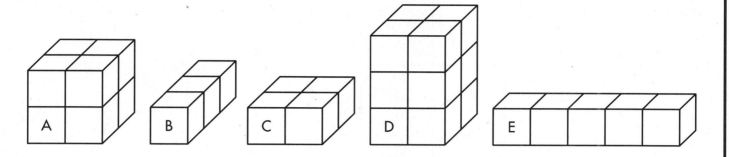

These five packages will be shipped in box 1. How many of each will fit in that box? (You may not break apart packages.)

Predict. Record your prediction. Then make the box and check your prediction. (Use the pattern on page 2). Record the actual number.

Box 1
4 by 6 cubes on the bottom and 3 cubes high

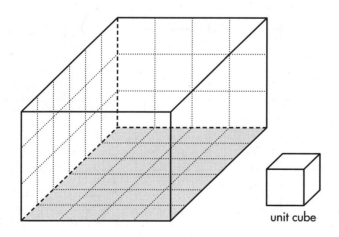

unit cube

How many of each package will fit in box 1?

	Prediction	Actual
A	_____	_____
B	_____	_____
C	_____	_____
D	_____	_____
E	_____	_____

How Many Packages? (page 2 of 4)

Pattern for Box 1

How Many Packages? (page 3 of 4)

Now work with packages A, D, and E and box 2. How many of each package will fit in this box? Predict, then make the box and check. (See page 4 for the pattern.) Record your predictions and your actual findings in the table.

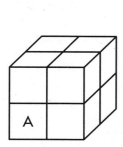

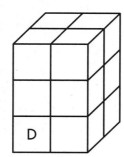

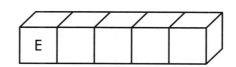

Box 2

4 by 6 cubes on the bottom and 5 cubes high

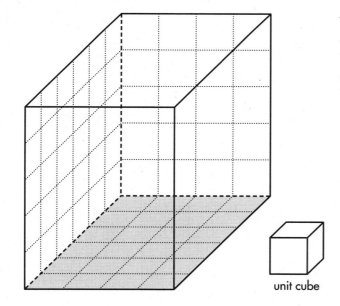

unit cube

How many of each package will fit in box 2?

	Prediction	Actual
A	_____	_____
D	_____	_____
E	_____	_____

How Many Packages? (page 4 of 4)

Pattern for Box 2 Add these strips to the sides of box 1 to make box 2.

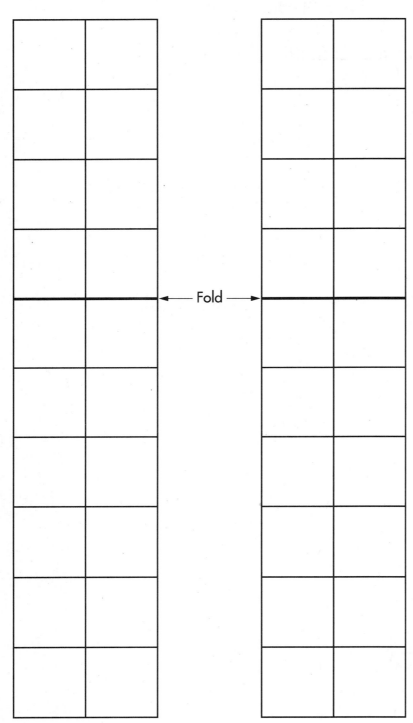

Fold

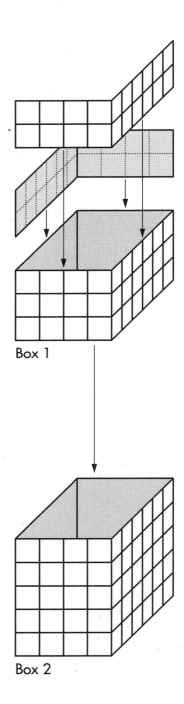

Box 1

Box 2

Containers and Cubes

Design a Box

Design a single open box so that packages of size A completely fill your box, packages of size B completely fill your box, packages of size C completely fill your box, and packages of size D completely fill your box. Be prepared to convince the class that your solution is correct.

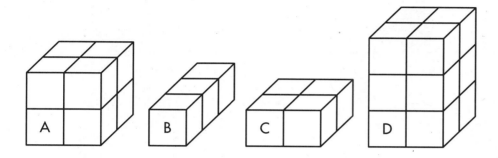

When you have made a box and tested that it works, record its dimensions.

Dimensions: _____
Can you find other boxes that will work? What are their dimensions?

Challenge

Design a single open box that can be completely filled with package A, or B, or C, or D, and also with package E. What are the dimensions of this box? Can you find other boxes that will work?

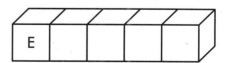

Dimensions: _____

More Packing Problems (page 1 of 5)

1. Predict how many packages made of two cubes it will take to completely fill box 1.
   Prediction: _____

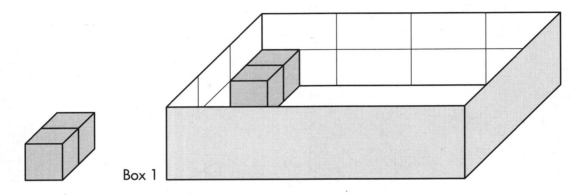

Box 1

   Describe how you made your prediction.

2. Make box 1. (Use the pattern on page 2.) Check your prediction with two-cube packages. What is the actual answer?

   Actual: _____

More Packing Problems (page 2 of 5)

Pattern for Box 1

More Packing Problems (page 3 of 5)

3. Predict how many packages made of two cubes it will take to completely fill box 2.

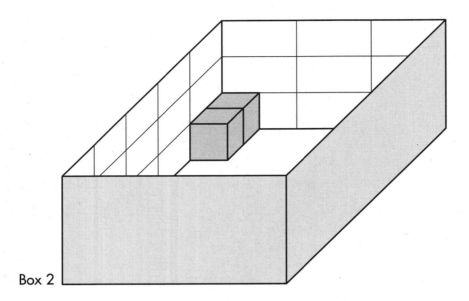

Box 2

Prediction: _____

Describe how you made your prediction.

4. Make box 2. (Use the patterns on pages 4 and 5.) Check your prediction with two-cube packages. What is the actual answer?

Actual: _____

More Packing Problems (page 4 of 5)

Pattern for Box 2

Cut out the right and left sides below. Cut out the rest of the pattern on the next page. Attach the sides to make the pattern for box 2, as shown in the diagram.

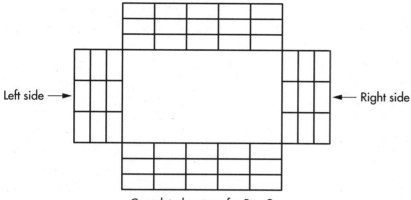

Completed pattern for Box 2

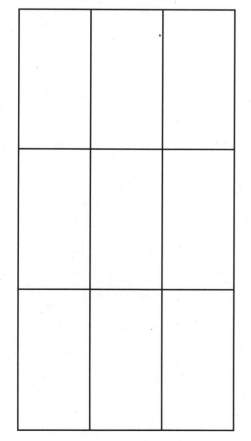

Left side

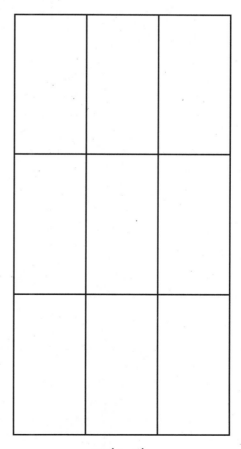

Right side

More Packing Problems (page 5 of 5)

Saving Cardboard

The boxes made by your company are built from cardboard. Below is the pattern for a box. Notice that this pattern includes the box top. How much cardboard is needed for this pattern? How many cubes will fit inside this box?

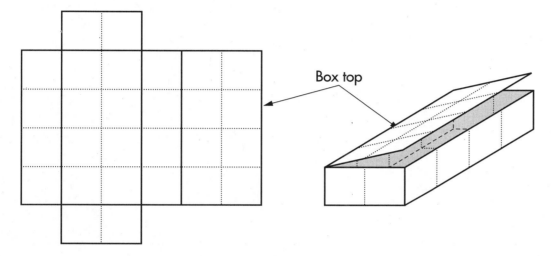

Box top

Your boss wants to save cardboard. Can you design a box that still holds 8 cubes but uses less cardboard than the box shown above? Find all possible boxes that hold 8 cubes, then choose the design that uses the least cardboard. Don't forget to include the box top in your design.

Pattern for a Closed Box

Cut out, fold, and
tape to make a
closed box.

How many cubic
centimeters will it
take to fill this box?
Find out without
filling the box with
cubes.

Pairs of Solids

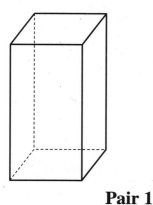

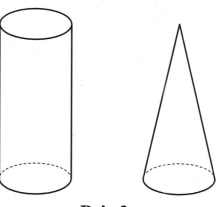

Pair 1

rectangular   rectangular
prism A   pyramid B

Pair 2

cylinder C   cone D

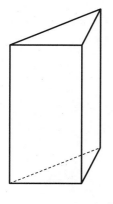

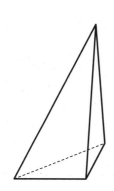

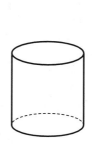

Pair 3

triangular   triangular
prism E   pyramid F

Pair 4

cylinder G   cone H

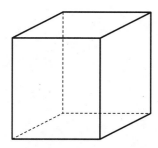

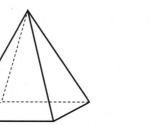

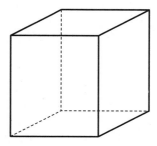

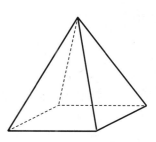

Pair 5

rectangular   rectangular
prism I   pyramid J

Pair 6

rectangular   rectangular
prism I   pyramid K

Pyramid and Prism Partners

Cut, fold, and tape to make a rectangular pyramid.

Use the grid below to design a rectangular prism with three times the volume of the pyramid.

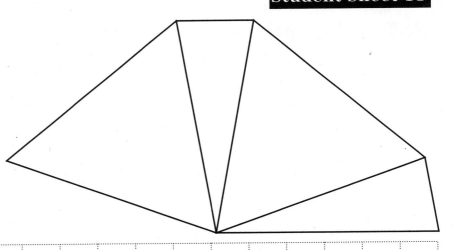

Puzzle Cube Pattern

Cut, fold, and tape to make an open cube. Show that this cube has three times the volume of the pyramid (Student Sheet 13).

Puzzle Pyramid Pattern

Cut, fold, and tape to make a pyramid. Show that this pyramid has one-third the volume of the open cube you made (Student Sheet 12).

Final Project Tasks

Rules

- Your model must have at least 8 and no more than 10 solids.

- Your model must have at least one pyramid, one prism, one cone, and one cylinder.

Tasks

1. Make a sketch of your model. Plan each solid you will need. Number the solids in your model as you record them on the Model Planning Sheet. Record their dimensions.

2. Make a pattern for each solid. Cut, fold, and tape the solids.

3. Predict the total volume of your model in cubic centimeters. Write your prediction on notebook paper. Tell in writing how you came up with your prediction.

4. Find the actual volume of your model. Write about how you did this. Record the volume of each solid on your Model Planning Sheet. Round volumes to the nearest whole cubic centimeter.

5. Use tape or glue to build your model.

6. Compile your final report. Include:
   - your sketch
   - your prediction of volume and how you got it
   - your completed Model Planning Sheet
   - your actual model

Model Planning Sheet

Description of your model:

Number	Name of solid	Dimensions	Volume

Total volume _____

On the back of this sheet, describe how you found the actual volume of each of your solids.

Solid Patterns A and B

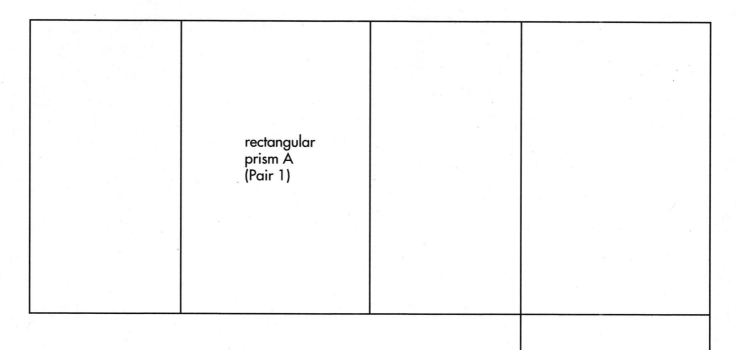

rectangular
prism A
(Pair 1)

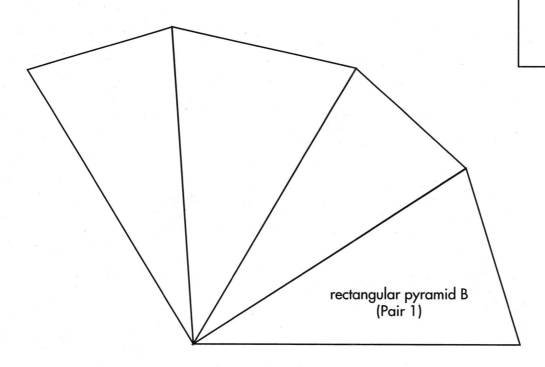

rectangular pyramid B
(Pair 1)

Solid Patterns C and D

cylinder C
(Pair 2)

cone D
(Pair 2)

Solid Patterns E and F

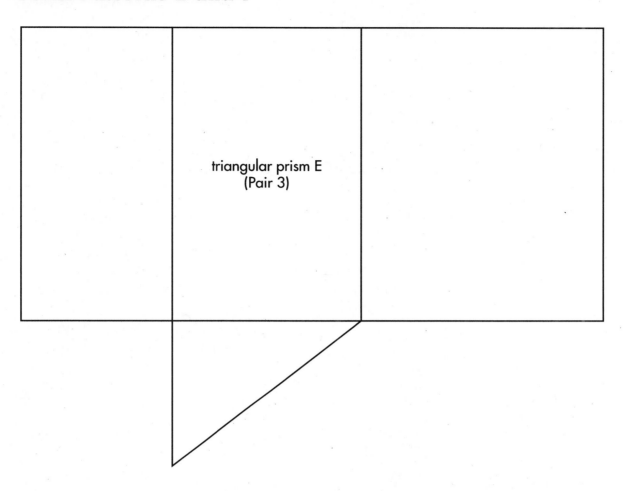

triangular prism E
(Pair 3)

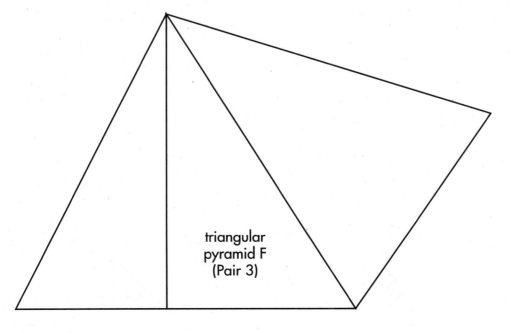

triangular
pyramid F
(Pair 3)

Fold the pyramid
so that its base
has the shape
of this
triangle.

Solid Patterns G, J, and K

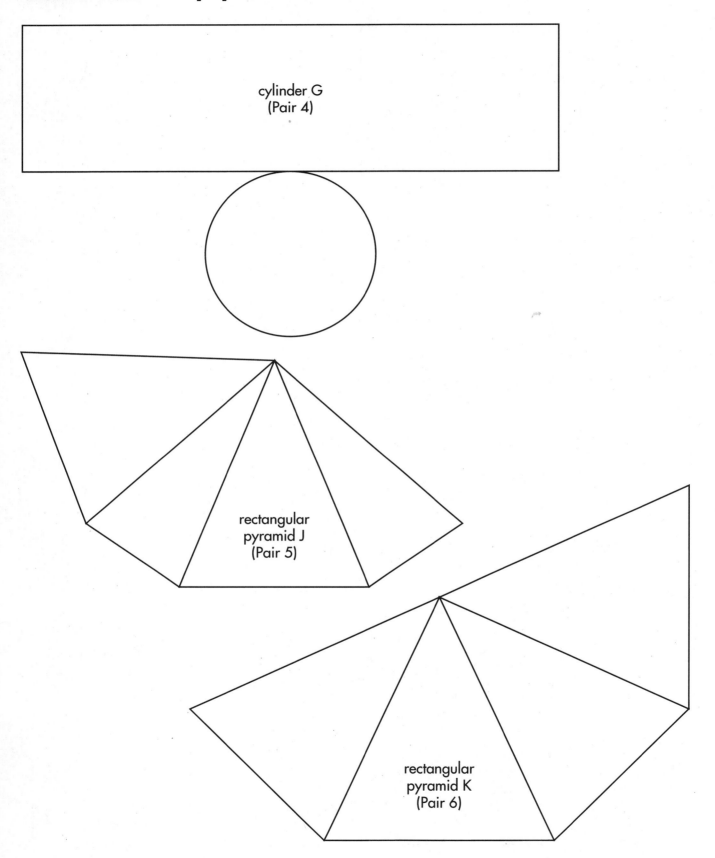

cylinder G
(Pair 4)

rectangular
pyramid J
(Pair 5)

rectangular
pyramid K
(Pair 6)

Solid Patterns H and I

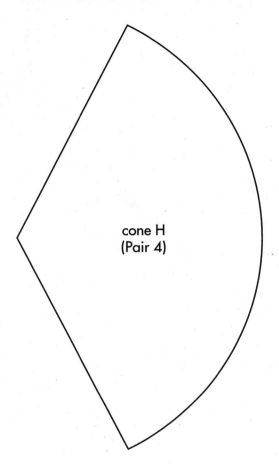

cone H
(Pair 4)

rectangular prism I
(Pairs 5 and 6)

Cone Pattern

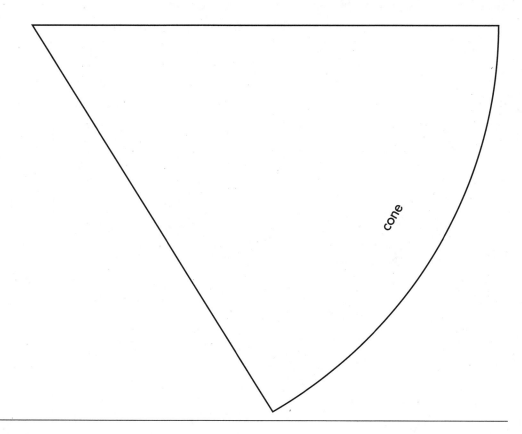

cone

Pyramid Pattern

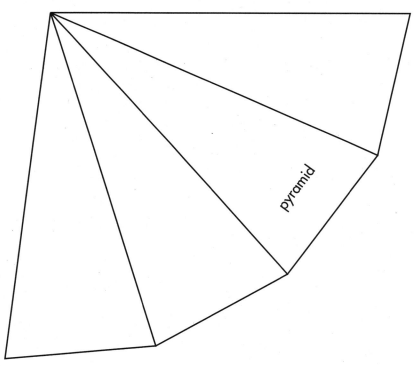

pyramid

Cut this page in half to distribute the cone and
pyramid patterns at different times.

Cylinder Pattern

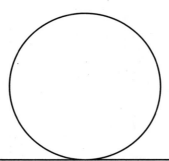

cylinder

See-through Graduated Prism Pattern

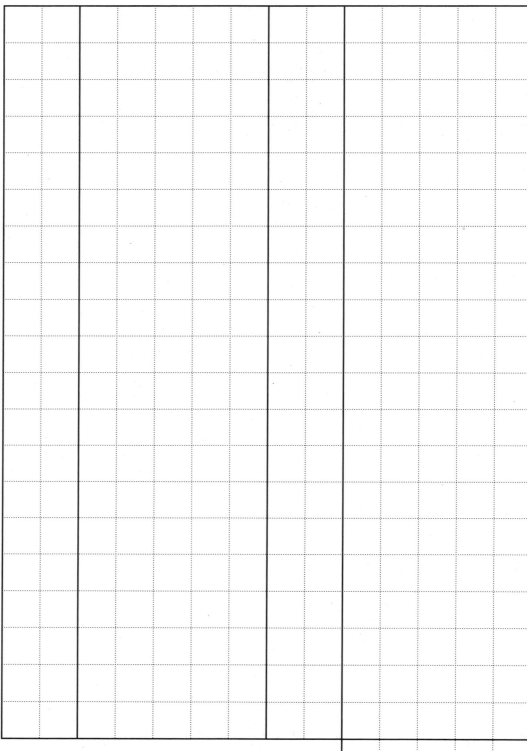

Use this if you don't have the plastic see-through prisms. Copy this pattern onto a sturdy overhead transparency. Cut out. Lay a ruler along each heavy fold line and crease the film over the ruler with your finger. Use transparent tape along the full length of the side edge and three base edges.

Dimensions of Solids

Pyramid

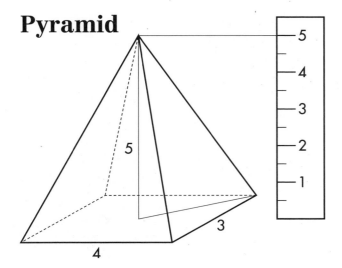

base: 4 by 3
height: 5

Cone

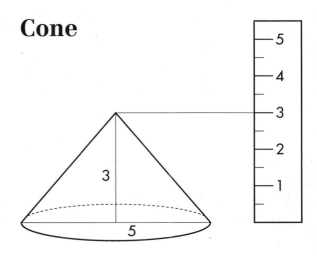

diameter of base: 5
height: 3

Prism

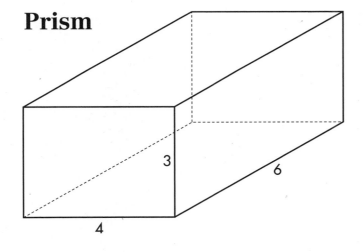

base: 4 by 6
height: 3

Cylinder

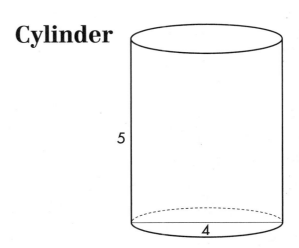

diameter of base: 4
height: 5

Cut along this line,
then along line
A, B, or C.

A makes a thinner cone.

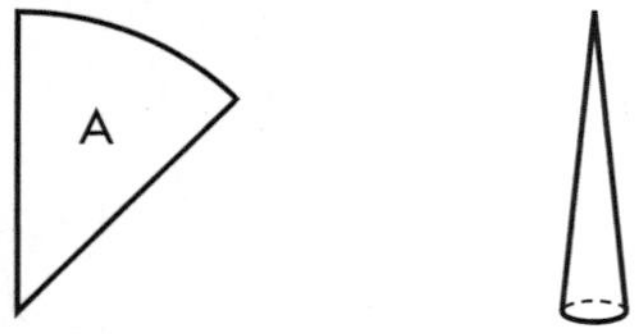

B makes an average cone.

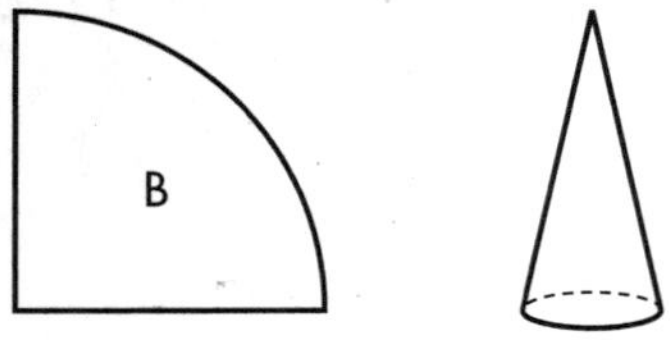

C makes a wider cone.

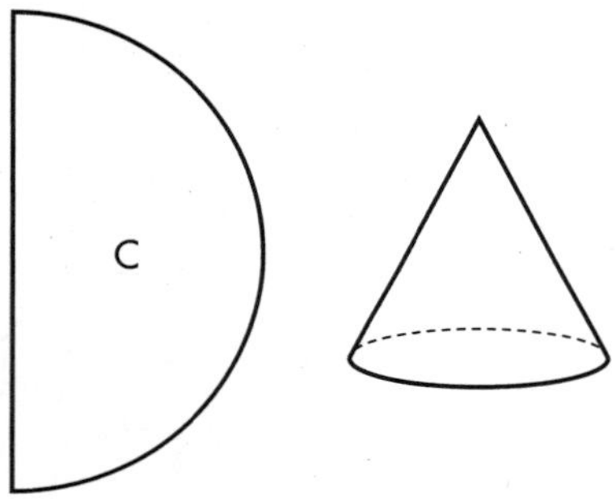

Cut along a curved line to
get the height that you want.
The number tells you the
diameter of the base, in
centimeters.

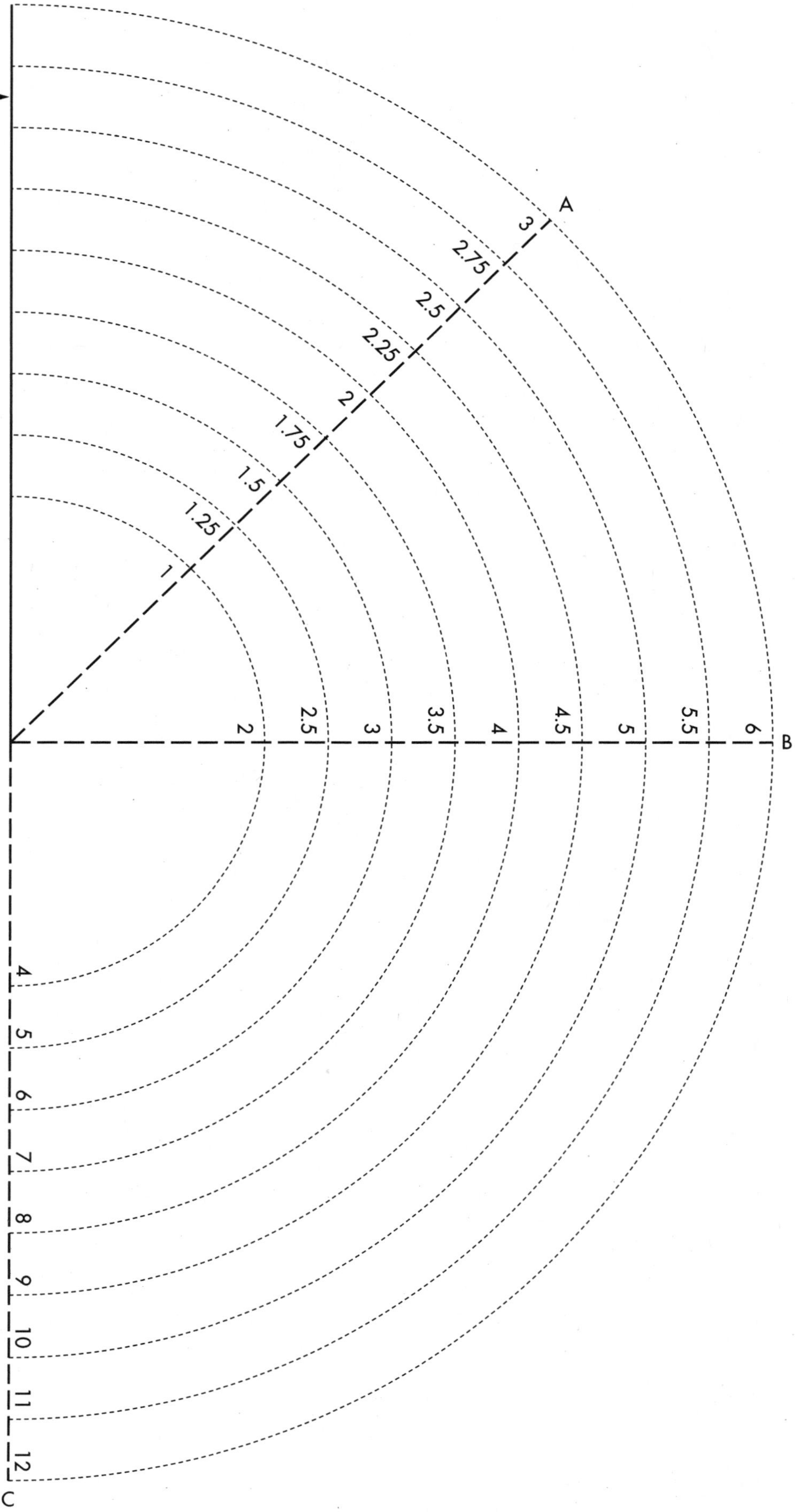

A

B

C

3
2.75
2.5
2.25
2
1.75
1.5
1.25

2
2.5
3
3.5
4
4.5
5
5.5
6

4
5
6
7
8
9
10
11
12

Cut out a base, then tape it to the corresponding cylinder side (see the next sheet).

2 cm

A

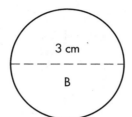

3 cm

B

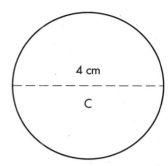

4 cm

C

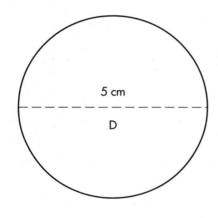

5 cm

D

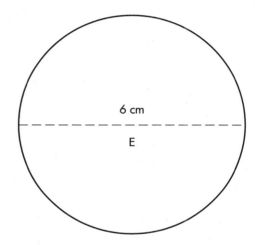

6 cm

E

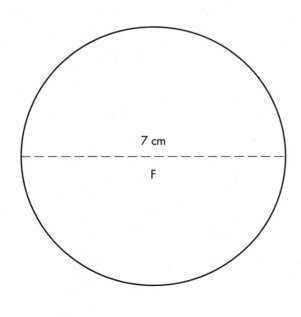

7 cm

F

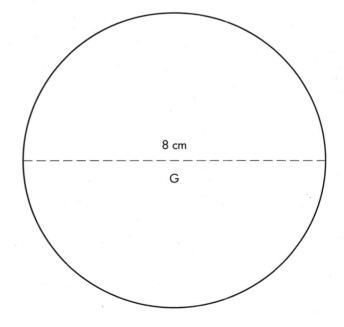

8 cm

G

Cut along a dashed line to get the side for the base you have chosen. For example, cut along line C if you want to make a cylinder with base C. Then cut along a dotted line to get the height that you want.

A	B	C	D	E	F	G

height in cm

2
3
4
5
6
7
8
9
10
11
12
13
14
15
16

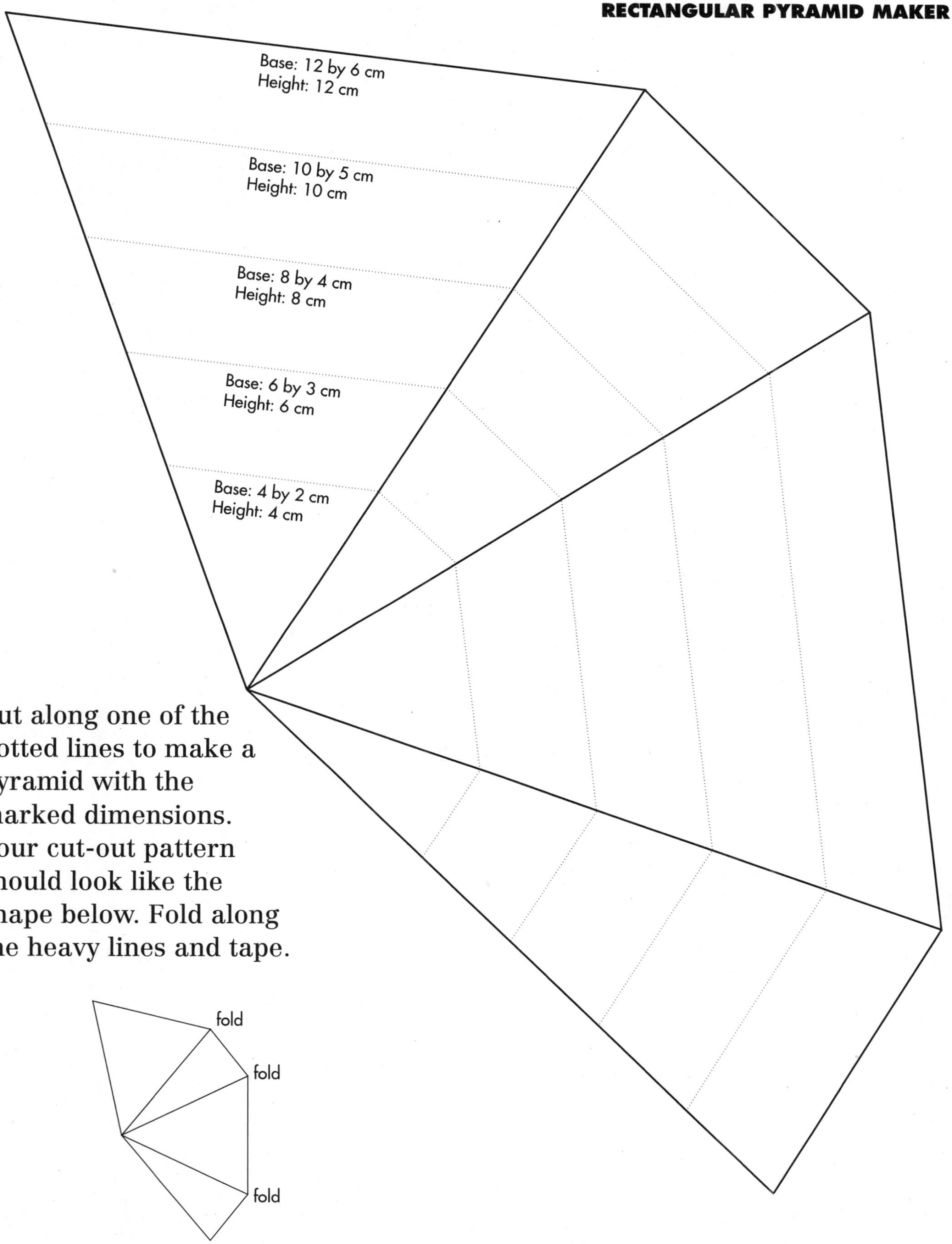

Base: 12 by 6 cm
Height: 12 cm

Base: 10 by 5 cm
Height: 10 cm

Base: 8 by 4 cm
Height: 8 cm

Base: 6 by 3 cm
Height: 6 cm

Base: 4 by 2 cm
Height: 4 cm

Cut along one of the dotted lines to make a pyramid with the marked dimensions. Your cut-out pattern should look like the shape below. Fold along the heavy lines and tape.

fold

fold

fold

Cut along one of the dotted lines to make a pyramid with the marked dimensions. Your cut-out pattern should look like the shape below. Fold along the heavy lines and tape.

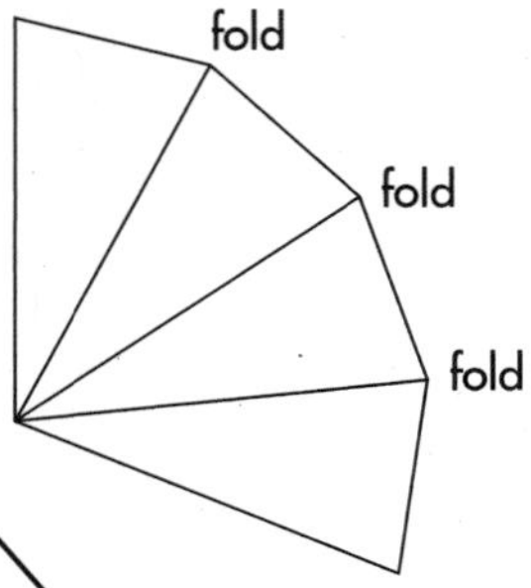

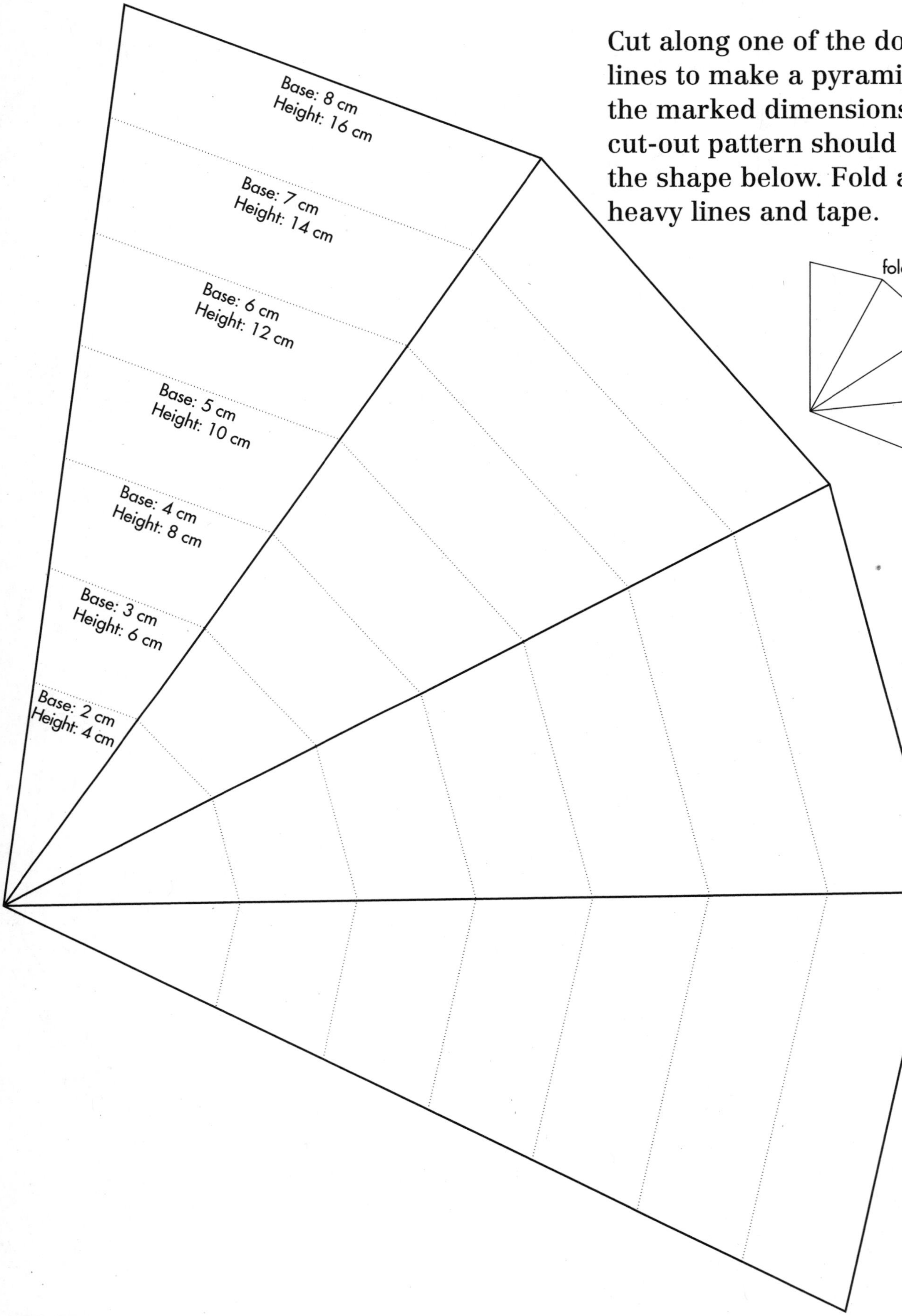

Base: 8 cm
Height: 16 cm

Base: 7 cm
Height: 14 cm

Base: 6 cm
Height: 12 cm

Base: 5 cm
Height: 10 cm

Base: 4 cm
Height: 8 cm

Base: 3 cm
Height: 6 cm

Base: 2 cm
Height: 4 cm

fold

fold

fold

Cut along a dotted line to get the height that you want. Fold along the heavy lines and tape, so the sides fit the bottom edges.

		15 cm
		14 cm
		13 cm
		12 cm
		11 cm
		10 cm
		9 cm
		8 cm
		7 cm
		6 cm
		5 cm
		4 cm
		3 cm
		2 cm

height

3 cm   6 cm

bottom

Guess My Unit Cards, Sheet 1

Duplicate a set of cards for each pair or small group, on card stock if possible. Use the blank cards to add other units of measurement that may have come up in class.

If you have students with limited English proficiency, help them identify each unit and draw a simple picture on the card that will help them identify that unit as they are playing the Guess My Unit game.

mile	foot	inch
meter	centimeter	millimeter
kilometer		

pound	ounce	ton
kilogram	gram	milligram

quart	cup	fluid ounce
liter	milliliter	cubic centimeter

second	minute	hour
day	week	month
year	decade	century
